T0282346

# Functional Diversity of Mycorrhiza and Sustainable Agriculture

# Functional Diversity of Mycorrhiza and Sustainable Agriculture

## Management to Overcome Biotic and Abiotic Stresses

**Michael J. Goss**
School of Environmental Sciences, University of Guelph,
Guelph, Ontario, Canada

**Mário Carvalho**
Institute of Mediterranean Agriculture and Environmental Sciences,
University of Évora, Évora, Portugal

**Isabel Brito**
Institute of Mediterranean Agriculture and Environmental Sciences,
University of Évora, Évora, Portugal

Academic Press is an imprint of Elsevier
125 London Wall, London EC2Y 5AS, United Kingdom
525 B Street, Suite 1800, San Diego, CA 92101-4495, United States
50 Hampshire Street, 5th Floor, Cambridge, MA 02139, United States
The Boulevard, Langford Lane, Kidlington, Oxford OX5 1GB, United Kingdom



**British Library Cataloguing-in-Publication Data**
A catalogue record for this book is available from the British Library

**Library of Congress Cataloging-in-Publication Data**
A catalog record for this book is available from the Library of Congress

ISBN: 978-0-12-804244-1

For Information on all Academic Press publications
visit our website at https://www.elsevier.com/books-and-journals



*Publisher:* Nikki Levy
*Acquisition Editor:* Nancy Maragioglio
*Editorial Project Manager:* Billie Jean Fernandez
*Production Project Manager:* Lisa Jones
*Cover Designer:* Mark Rogers

Typeset by MPS Limited, Chennai, India

# Contents

## 4. Diversity in Arbuscular Mycorrhizal Fungi

*With Clarisse Brígido*



## 5. Impacts on Host Plants of Interactions Between AMF and Other Soil Organisms in the Rhizosphere

*With Luís Alho and Sabaruddin Kadir*



## 6. The Significance of an Intact Extraradical Mycelium and Early Root Colonization in Managing Arbuscular Mycorrhizal Fungi

# List of Figures

# List of Plates

# List of Tables

# Preface

The current world population of 7.5 billion is expected to be 20% greater by 2050 and so we have little over 33 years to ensure the means of producing sufficient food to meet the expected demand. One of the options that previously were available to us for expanding world production of cereals, vegetables, fruits, and meat, namely bringing more land into production, is no longer possible and consequently we must everywhere increase the productivity of the land. But this time we must not attempt it without making every effort to safeguard the environment. Put in a slightly different way, we have to grow more but conserve the soil and its biodiversity, be more efficient in terms of water use, improve nutrient-use efficiency so that fewer applied nutrients end up contaminating our freshwater and eutrophying our lakes and shallow seas or adversely affecting the quality of our air and contributing to the atmospheric loading of greenhouse gases. If we add in a desire to reduce the application of pesticides, especially those targeting root pathogens, it would seem to represent an extremely challenging task. Perhaps it will be a surprise to some that the answer to many of these challenges might well be one result of the development of techniques that allow us to determine the make-up of microorganisms, which has had huge impacts on soil science and its application in agronomy.

Beginning with the ability to differentiate the fatty acid and phospholipid profiles of microbial communities in soil and reaching the current status, where the whole genetic code of an organism can be determined, the previously rather opaque world of soil microbiology is being clarified at an unprecedented rate. From around the time that the word mycorrhiza was coined by Frank in 1885, mycorrhizal fungi have been of interest because of their special relationship with the vast majority of land plants. For agronomists the most important are the endomycorrhizal fungi that produce tree-shaped branched structures called arbuscules inside the cortex of most crop plants. Evidence steadily accrued that established their importance in supplying the essential element phosphorus to plants but the availability of mineral fertilizers, such as superphosphate, caused many to assume that the contribution from mycorrhiza was unnecessary and even in fertile soils the organisms were more like parasites than partners of their hosts. But eventually there came the realization that arbuscular mycorrhiza provided far more services than supplying phosphorus. The recent appreciation of the biological diversity of mycorrhizal fungi and the functional consequences for mycorrhiza with different abilities to protect their host from the impacts of toxic metals, to counter the invasion of root diseases and to enhance the formation and stabilization of soil aggregates, renewed interest in the ecological significance of mycorrhiza.

The challenge for agronomists and those interested in availing their crops of the potential benefits from arbuscular mycorrhiza is how to manage them. One obvious approach is to develop a source of inoculum that can be applied to a field prior to or as part of seeding a crop that could benefit from the formation of a mycorrhizal symbiosis. However, not only is that a relatively expensive activity it is also fraught with uncertainty over its efficacy. Another approach is to encourage the adoption of farming practices that support a wide variety of indigenous arbuscular mycorrhizal fungi (AMF) that will provide specific benefits sought for the crop. But in many respects this is not enough. It is a long way from providing the supportive environment for a specific fungus or consortium of fungi to dominate the mycorrhiza that form on most crop plants in a field. That goal requires the development of new farming strategies.

The approach we take in this book is to expand on the current challenges to meeting the requirements for feeding a much larger world population and suggest how arbuscular mycorrhiza can contribute to the solution under many agricultural climatic zones. We consider the farming practices that can be deleterious to maintaining a diverse population of mycorrhizal fungi and the systems and practices that can encourage their survival and effectiveness. We discuss the interactions between the fungi and other soil organisms, some of which are now known to improve the functioning of arbuscular mycorrhiza, and how the symbiosis influences many of the basic plant processes. The possibilities for obtaining specific information on individual fungi offered by the new generation of molecular methods are also presented. Finally we present a view as to how indigenous AMF might be managed in a practical setting.

The opportunity to put our combined thoughts and ideas into a book owes a lot to the discussions we had with Marisa LaFleur, commissioning editor with Elsevier, and subsequently with commissioning editor Nancy Maragioglio. Both have been wonderfully supportive and we can't thank them enough. We are equally indebted to Billie Jean Fernandez, who has been of enormous help in pulling us over the finish line. Lisa Jones, the Production Editor, has been superb in converting our ideas on presentation into reality; she has worked tirelessly to ensure we would be proud of the finished product. We sought the help of two experts to ensure that the chapters on new generation molecular methods and diversity among the AMF would be as up-to-date as possible. It is difficult to express just how grateful we are to Diederik van Tuinen, a very good friend and colleague, for his contribution on modern molecular methods in relation to the elaboration of functional diversity. The contribution of Clarisse Brígido in developing the chapter discussing the complexity of functional diversity in AMF was also critical and she too has been of incalculable help and support. We are extremely grateful to Sabaruddin Kadir and Luis Alho, who generously provided material used in Chapter 5, as well as provided important feedback on the contents.

**Michael J. Goss, Mário Carvalho, and Isabel Brito**
**March 2017**

# Taxonomy of Arbuscular Mycorrhizal Fungi Referred to in this Book

There have been some major changes in the taxonomy associated with arbuscular mycorrhizal fungi (AMF). In consequence, some have undergone more than one name change in the last 30 years. To avoid as much confusion as possible, the names used in the text are those reported by the authors of the papers referenced. We have used Schüßler and Walker (2010) and Redecker et al. (2013) to provide a list of the current names of these species.

| **Former Name** | **Genera** | **Specific Epithet** |
|---|---|---|
| *Acaulospora leavis* | *Acaulospora* | *leavis* |
| *Acaulospora morrowiae* | *Acaulospora* | *morrowiae* |
| *Entrophospora schenckii* | *Archaeospora* | *schenckii* |
| *Gigaspora albida* | *Gigaspora* | *albida* |
| *Gigaspora gigantea* | *Gigaspora* | *gigantea* |
| *Gigaspora margarita* | *Gigaspora* | *margarita* |
| *Gigaspora rosea* | *Gigaspora* | *rosea* |
| *Glomus caledonium (Nicol. & Gerd.) Trappe and Gerdemann* | *Funneliformis* | *caledonium* |
| *Glomus claroideum* | *Claroideoglomus* | *claroideum* |
| *Glomus clarum* | *Rhizophagus* | *clarus* |
| *Glomus constrictum = Funneliformis constrictum* | *Septoglomus* | *constrictum* |
| *Glomus coronatum* | *Funneliformis* | *coronatus* |
| *Glomus diaphanum* | *Rhizophagus* | *diaphanus* |
| *Glomus etunicatum* | *Claroideoglomus* | *etunicatum* |
| *Glomus fasciculatum* | *Rhizophagus* | *fasciculatus* |
| *Glomus fasciculatum Gerd. And Trap* | *Rhizophagus* | *fasciculatus* |
| *Glomus fasciculatum (Thaxter sensu Gerd)* | *Rhizophagus* | *fasciculatus* |
| *Glomus geosporum* | *Funneliformis* | *geosporum* |

(Continued)

(Continued)

| Former Name | Genera | Specific Epithet |
|---|---|---|
| *Glomus intraradices*[a] | *Rhizophagus* | sp. |
| *Glomus intraradices* | *Rhizophagus* | *irregularis* |
| *Glomus intraradices* | *Rhizophagus* | *intraradices* |
| *Glomus macrocarpum* | *Glomus* | *macrocarpum* |
| *Glomus mosseae* | *Funneliformis* | *mosseae* |
| *Glomus tenue* | *Glomus* | *tenue* |
| *Rhizophagus intraradices* | *Rhizophagus* | *intraradices* |
| *Scutellospora calospora* | *Scutellospora* | *calospora* |
| *Scutellospora fulgida* | *Racocetra* | *fulgida* |

[a]Identifying the current name for *Glomus intraradices* is problematic. The isolate DAOM197198 was renamed from *Glomus intraradices* to *Glomus irregularis* and then to *Rhizophagus irregularis.* As not all isolates have been reanalyzed, we now have some which are *Rhizophagus* sp., some *R. irregularis,* and some still *R. intraradices.*

# Chapter 1

# Challenges to Agriculture Systems

## Chapter Outline



## 1.1 CURRENT AND FUTURE CHALLENGES TO AGRICULTURE SYSTEMS

Food production is probably one of the greatest challenges facing the world. Despite the increase in agricultural production since the 1960s, when the "green revolution" started to be implemented in the developing world, we still have more than 1 billion undernourished people (FAO, 2009, 2015a). There has to be a greatly increased production simply to feed a population growing from 7 billion to in excess of 9 billion over the next 35 years (Fig. 1.1).

This growth in population, the improvement of world gross product (WGP) and consequent greater consumption of food, together with changes to the human diet, particularly the switch to grain-fed animal protein, all combine to exert further pressure on agricultural production. Even allowing for the uncertainties related to each of these factors, it is estimated that by the year 2050 world food production will have to increase by 50%–70% (FAO, 2009; The Royal Society, 2009).

A key concern is how this additional production is going to be achieved. In the past, the response in both developed and developing countries to a greater demand for food has been to increase the area made available for agriculture and enhancing land productivity by an increase in crop yields. For example, over the period 1961–2005, expansion of harvested land contributed between 14% and 25% to improved crop production compared with the 78%–86% resulting from improved productivity, with about 10% of the latter resulting from increased cropping intensity, that is the ratio of harvested land to the total arable land (Table 1.1) (Bruinsma, 2011).

Functional Diversity of Mycorrhiza and Sustainable Agriculture.
DOI: http://dx.doi.org/10.1016/B978-0-12-804244-1.00001-0

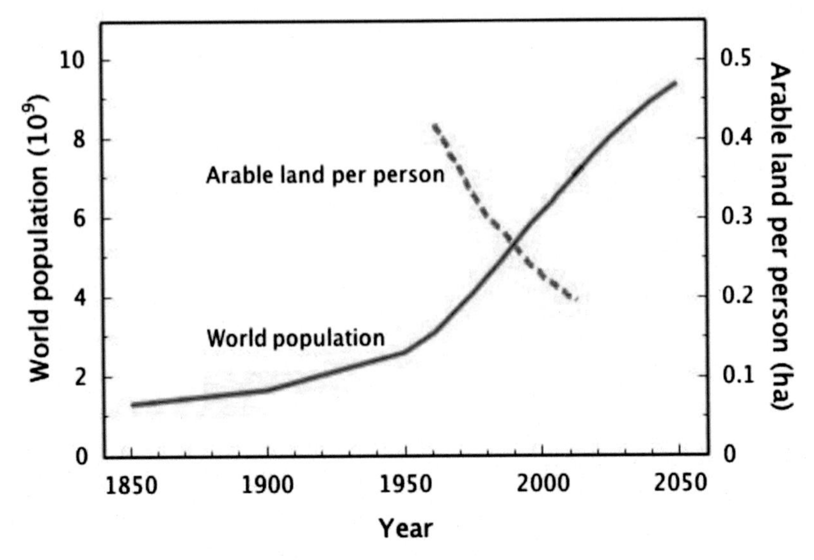


**FIGURE 1.1** The rapid increase in world population since 1960 and the associated reduction in the average area of arable land per person. Note that the average area of arable land area per person is less than half that in 1960 and is now smaller than 0.2 ha (FAOSTAT, 2015; US Census Bureau, 2014).

**TABLE 1.1** Estimation of relative contributions to improved crop production of increases in harvested land area, crop yields[a] and cropping intensity of agriculture over the period from 1961 to 2005 (Bruinsma, 2011).

| | Harvested Area (%) | Crop Yields (%) | Cropping Intensity (%) |
|---|---|---|---|
| Developing countries | 22 | 70 | 8 |
| World | 14 | 77 | 9 |

[a]*Weighted yields (international price weights) based on 34 crops.*

Nevertheless, the aggregate land area in developed countries showed a decline over the same period, so improvement in yield was even more important as a factor (Bruinsma, 2003).

The available evidence strongly points to the conclusion that increasing the land area under cultivation will be inadequate as an option to meet the current challenge. In 30 years after 1950, more land was converted to cropland than in the 150 years between 1700 and 1850 (Millennium Ecosystem

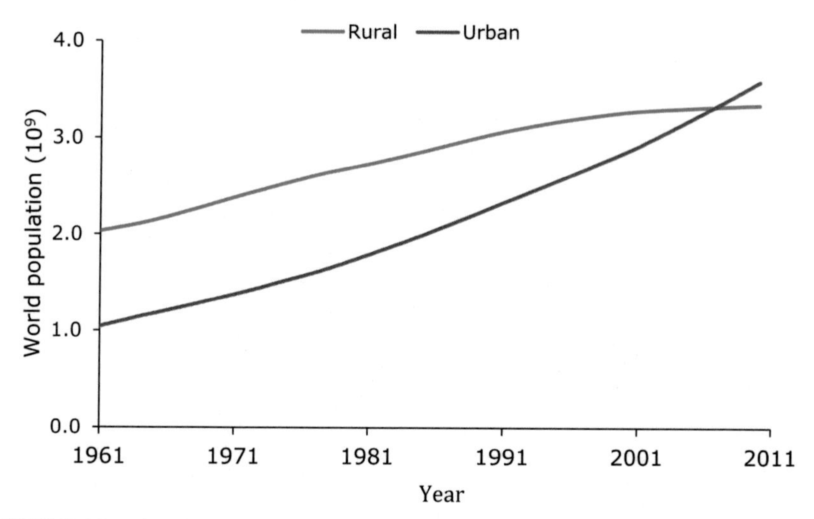


**FIGURE 1.2** The urbanization of the world population since 1961 (FAOSTAT, 2015).

Assessment, 2005). Worldwide agriculture has already been responsible for the conversion of 70% of grassland, 50% of savanna, 45% of temperate deciduous forest, and 27% of tropical forest biome (Foley et al., 2011). This represents 38% of Earth's terrestrial surface with the soils most suitable for agriculture already under cultivation. In addition to the reduced opportunity for further land use change, good agricultural land is lost every year to build houses and necessary infrastructures to accommodate the growth of the world population and the migration to cities (Fig. 1.2). By 2030 there are expected to be 1.75 billion more urban residents, requiring about 42.4 million ha of new urban land cover (Dumanski, 2015).

Degradation of land due to desertification, soil erosion from water or wind, acidification, nutrient deficiency or being affected by salt, compaction, or contamination by toxic materials is also a threat to the available land dedicated to food production (Box 1.1). In addition to these aspects, climate change will impact land productivity in large area of the planet and will likely compound the negative impacts of agriculture on land degradation. The net combination of anthropogenic impacts has meant that the average area of arable land per person has more than halved since 1960 and is currently a little less than 0.2 ha (Fig. 1.1).

*If a significant increase in the land area dedicated to agriculture is not an option then, greater soil productivity is an essential strategy to face the challenge of feeding the world population in the near future.*

These concerns result in a need to improve production (better postharvest preservation is insufficient) but in a context of reduced manufactured inputs and restricting impacts on the environment. In particular, increased production needs to take place where inputs are traditionally much less or are even not available.

The impact of the conversion of forest and native grassland to agricultural production on biodiversity at the planet scale has been tremendous and, across a range of taxonomic groups, the population size or range (or both) of the majority of species is declining (Millennium Ecosystem Assessment, 2005). Because the most fertile areas have long been exploited, any further increase of the land dedicated to agricultural production will be at the expense of high value ecological reserves, especially important for biodiversity, carbon storage, and water cycling but with less proportional benefits in terms of their potential for food security. The consequences for the production of food crops is that increasingly they will be grown on less favorable land that has poorer soil quality and is more susceptible to degradation.

The technological advances in agricultural production, based on crop breeding and an increased use of inputs, such as fertilizers, pesticides, and irrigation, has resulted in an increase of crop yields, which has been responsible for nearly three-quarters of the enhancement of production since 1961 (Dixon et al., 2001, Table 1.1). However, adverse effects on the wider environment, such as water pollution and scarcity, land degradation, loss of biodiversity, and increase in anthropogenic greenhouse gas emission have been the counterpart of this increase in productivity. In the past 50 years the world's irrigated cropland area roughly doubled, while global fertilizer use increased by 500% (Foley et al., 2011). However, the benefits were not evenly distributed across all crops or world regions and impacts of the intensification of agriculture on the environment also differ across the world. Cereals and developed countries have shown the largest yield increases, whereas many of the staple crops in the tropics together with minor cash crops have received relatively little attention from the private research carried out by the major seed companies (Dixon et al., 2001). China has been very successful in raising yields in the last 50 years with an increase in food production per capita of almost 3.5 times (Godfray et al., 2010). In contrast in Sub-Saharan Africa (SSA) and parts of South Asia people are still suffering chronic undernourishment. Per capita production fell back from the mid-1970s in SSA (Godfray et al., 2010). Nitrogen use per person in 2001 was only 1.1 kg in SSA compared to 22 kg in China and 38 kg in the United States (Mosier et al., 2004). In SSA annual NPK nutrient balances are still negative (Stoorvogel et al., 1993) and probably with an increased tendency for further decline in the near future.

**BOX 1.1 Soil Degradation**

Soil degradation is defined as a change in the soil quality status resulting in a diminished capacity of the ecosystem to provide goods and services for its beneficiaries (http://www.fao.org/soils-portal/soil-degradation-restoration/en/). It is estimated that 5 billion hectares are degraded worldwide, with 64% of this area in dry regions (Eswaran et al., 2001). There are several causes for land degradation. Erosion by water and wind is the main cause and contributes to about 85% of land degradation (Oldeman et al., 1992). On a global scale the costs to the world of an annual loss of 75 billion tonnes of soil is about US$ 400 billion year$^{-1}$, or approximately US$ 70 person$^{-1}$ year$^{-1}$ (Pimentel et al., 1995; Lal, 1998). Soil compaction is mainly important in the regions of the world where mechanization has been intensively used. On-farm losses through land compaction in the United States have been estimated at US$ 1.2 billion year$^{-1}$ (Gill, 1971), and it has caused yield reductions of 25%–50% in some regions of Europe (Eriksson et al., 1974) and North America, and between 40% and 90% in West African countries (Charreu, 1972; Kayombo and Lal, 1994). Soil acidity also threatens crop yields, either by reducing the availability of important nutrients for crop nutrition or through the associated toxicities of Al and Mn. Around 50% of the world's potentially arable soils are acidic and the use of fertilizers and biological nitrogen fixation are promoting soil acidity. Salinization is also an important aspect of soil degradation. Salt-affected soils occur in more than 100 countries and their worldwide extent is estimated at about 1 billion ha (FAO and ITPS, 2015). Some 10%–20% of dry lands are already degraded due to desertification while a much larger number is under threat (Millennium Ecosystem Assessment, 2005). The United Nations Environment Programme (UNEP) estimates that 16% of the world's productive land is already degraded (Parry et al., 2009).

## 1.2 THE APPROACH TO MEETING THE CHALLENGES TO WORLD AGRICULTURE

*To feed the world population and simultaneously reduce environmental impacts at the global scale is not just a question of technological advances. There is a need for economic and social action on a global scale.*

From a scientific viewpoint it is obvious that the current technology is not able to fully meet the challenge. Assuming that a significant increase of the area dedicated to agriculture is not an option, there is an urgent need to close the gap between the potential yield and the yield that farmers achieve in each agroecosystem across the different regions of the world. Each yield gap results from a single or multiple biotic and abiotic stresses that are not appropriately managed by the cropping system. The most significant cause of yield loss varies within and between different regions of the world. At a global scale nutrient deficiency and water stress are certainly major constraints to reaching the yield potential for any crop. Shortage or unbalanced nutrient application is evident in large regions of Eastern Europe, SSA, and South Asia, and consequently increased use of fertilizers will be one requirement to close yield gaps. However, there are other regions of the world where currently there is an excessive amount of fertilizer being applied. Mueller et al. (2012) project that smaller net changes in nutrient inputs will be required: 9%, 22%, and 34% modifications in the application of nitrogen (N), phosphorus (P), and potassium (K), respectively, to close yield gaps to 75% of attainable yields if fertilizers are used more efficiently at a global scale. Water stress limits yield in most regions of the world and water availability for irrigation is already under a great pressure. Therefore a combination of plant breeding and enhanced management techniques aimed at improving water use efficiency (WUE) will be crucial in closing those yield gaps that are due to water stress. Toxicity of both Al and Mn imposes an important limitation to crop yield mainly in the wetter regions of the world and liming will be an important component of the strategy needed to improve crop yields in those regions. Biotic stresses are also globally present and locally adapted strategies involving plant breeding together with integrated pest and weed management are also necessary to close the yield gap around the world under the overarching perspective of sustainable agriculture. This approach requires the evolution of what is usually described as the sustainable intensification of production systems (Box 1.2).

Given the economic, social, and environmental variations across the globe, different approaches and solutions will be needed and all scientific and technological options have to be considered without any ideological preconception. It is not possible to rule out the contribution from genetic manipulation, use of inputs such as fertilizers, pesticides, or irrigation, and from production systems, such as conservation and organic agriculture. Tradeoffs between different goals will have always to be considered, but short-term objectives cannot be allowed to compromise protection of soil and water, the maintenance of biodiversity at local and global scale and the mitigation and adaptation to climate change. Increase in the efficiency of resource use is crucial if real improvement is to be achieved.

There is a widespread deficiency of N and P and it is not possible to conceive sustainable crop intensification without the use of fertilizers. The use of other nutrients will have also to be considered, particularly K, although its need is more depended on the region and crop than is that for N and P. The nutrient use efficiency (NutUE), defined as the recovered rate in farm products of nutrients applied to the system (including fertilizers, manures, biological N fixation), is a crucial aspect of sustainable production systems, particularly that of N, P, and K. N and P are major contaminants of water,

**BOX 1.2 Sustainable Intensification of Agriculture**

*Increasing* world food production while protecting the environment is the new paradigm that current agricultural production systems have to follow and for which the concept of sustainable agriculture was developed. The intensification of agricultural production promoted by the technologies developed in the second half of the last century, particularly increased use of inputs, such as fertilizers and pesticides together with plant breeding, was very effective in increasing land productivity. However, the use efficiency of different production inputs was not improved, which is a major reason for the impacts of agriculture on the environment. Over the period between 1960 and 1990 the use of fertilizers and pesticides increased 4.8 and 31.1 times, respectively, whereas world production of the 20 most important agricultural commodities only doubled (Fig. B1.2.1). Sustainable intensification of agriculture is a concept and not a production system in own right. It is not a tool that can or needs to be incorporated within the various agricultural production systems practiced in different regions of the world. The aim is very challenging and achieving it requires a multidisciplinary and holistic approach. A contribution from many different technologies is an important requirement. Furthermore, a systematic and long-term approach is needed.

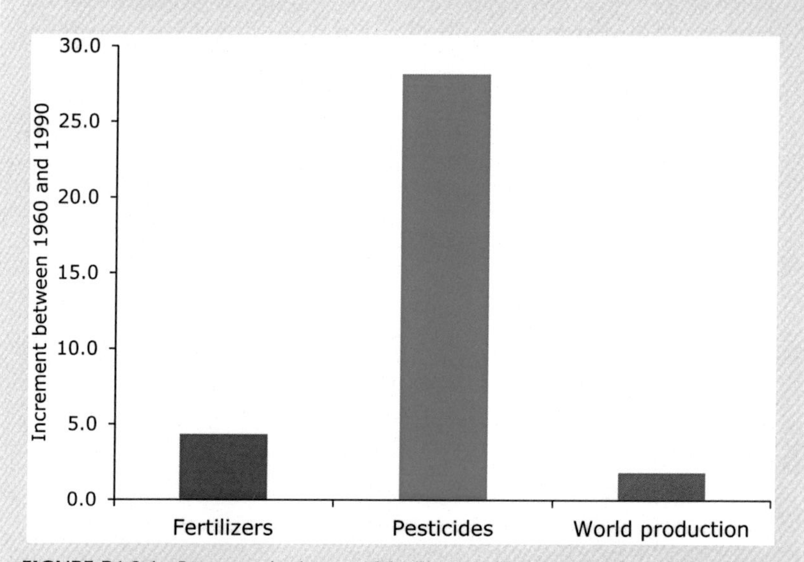


**FIGURE B1.2.1** Increment in the use of fertilizers (million tonnes of nutrients; adapted from Bumb and Baanante, 1996), pesticides (million US$, adapted from Zhang et al., 2011), and production (million tonnes of 20 most valuable commodities, FAOSTAT 2015) in the world between 1960 and 1990.

(*Continued*)

**BOX 1.2 (Continued)**

For instance it is not possible to improve the fertilizer use efficiency if abiotic (water deficiency, excessive levels of metal ions or salt, for example) or biotic stresses (weed competition or pests and diseases) are not taken into consideration. A basic and probably the most relevant change to be considered within cropping systems is the improvement of soil quality in relation to its reserves of nutrients, its water relations (drainage and water holding capacity) and overall quality. The most relevant soil parameter in relation to quality is the organic matter content. The first action of any sustainable intensification must be the control of soil erosion as topsoil contains the most organic matter and plant nutrients. During the last 40 years of the last century one-third of the world's arable land was lost by erosion and continuous to be lost at a rate of 10 million ha year$^{-1}$ (Pimentel et al., 1995). Soil erosion reduces soil organic matter (SOM) content, water-holding capacity, infiltration rates and soil biological activity, thereby affecting the efficiency with which production inputs are used. In consequence more inputs are needed and land productivity is diminished. Pimentel et al. (1995) estimated that investment in soil conservation techniques in the United States would have a return of $5.24 for each dollar invested. Binaj et al. (2014) estimated that 5.5% of agricultural GDP is lost annually due to soil degradation by erosion and compaction in Albania. Kefi and Yoshino (2010) estimated that for one watershed in Tunisia the reduction in farmer's income could reach 32% due to water erosion. For a watershed in Indonesia, Anasiru et al. (2013) estimated annual nutrient losses from erosion of 2648 kg C ha$^{-1}$, 230 kg N ha$^{-1}$, 30.5 kg P ha$^{-1}$ and 69 kg K ha$^{-1}$ corresponding to 14,231,904 IDR ha$^{-1}$ season$^{-1}$ (around US$ 904).

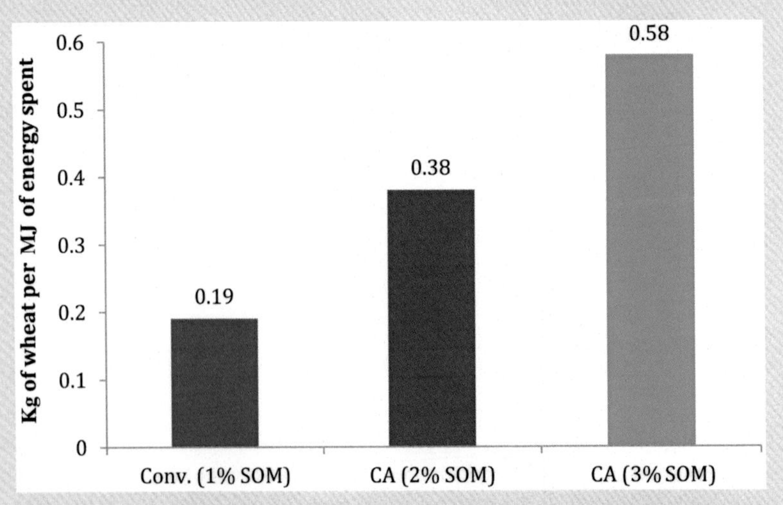


**FIGURE B1.2.2** Effect of soil organic matter (SOM) on the use efficiency of inputs to wheat production in the South of Portugal. *Conv.*, conventional agriculture system based on inversion tillage and bailing of the cereal straw; *CA*, conservation agriculture based on no-till and maintenance of crop residues on soil surface. The same crop rotation was used for the two systems.

(*Continued*)

**BOX 1.2 (Continued)**

Semalulu et al. (2014) estimated losses of US$ 172 $ha^{-1}$ $year^{-1}$ in annual crops in Uganda due to nutrient losses from erosion.

Conservation agriculture, based on no-till, the permanent cover of soil surface by crops and their harvest residues and crop rotation, is an efficient system to protect the soil against erosion and to improve SOM (Lal, 2015). Conservation agriculture has already contributed to the sustainable intensification of agriculture in several developed and developing regions of the world by improving soil quality and productivity but reducing the level of inputs (power, labor, fertilizers, agrochemicals, and water) needed; the extent of adoption worldwide is around 124,794,840 ha (Kassam and Brammer, 2013). After 10 years of adoption of conservation agriculture under Mediterranean conditions SOM increased and the use efficiency of inputs (transformed into equivalent energy for production and application) increased accordingly (Fig. B1.2.2) (Carvalho, 2013). The better use efficiency of the conservation agriculture systems resulted from the combined effect of greater land productivity (25%) and a reduction of energy inputs due to less power and fertilizers needed.

contributing to nitrate pollution of groundwater and eutrophication of superficial water reservoirs and shallow marine environments. The industrial production of N requires vast amounts of energy and water, and reserves of K and P in the world are decreasing. Fertilizers should not be applied if the reserves in the soil are creating environmental contamination even when the potential yield is achieved. When the soil reserves are only sufficient to support the potential yield in that season, the nutrient supplied needs to compensate for the output of the system to maintain optimum soil fertility. However, for most situations, and particularly for the crop nutrients N and P, soil reserves are insufficient to guarantee that a crop will attain its yield potential, so nutrients have to be applied. Because NutUE is relatively small, some loss of nutrients from the system is inevitable, as are the consequent environmental impacts. As an example, N efficiency rates vary from 23% in Africa to 52% in South America and for a European dairy herd, it is unusual to find recoveries that are more than 25%–30% in the meat or milk products of the farm (Goulding et al., 2008). Progress can be achieved in increasing NutUE by plant breeding, including the use of DNA transformation to generate genetically modified plants (GM), better time of N application, managing soil structure, and the development of slow release fertilizers. Increasing SOM content is an ideal way to improve the N efficiency rate, which under Mediterranean conditions can be doubled by increasing SOM from 1% to 2% (Carvalho and Lourenço, 2014). The contribution of biologically fixed nitrogen and the application of manure are also important as means of reducing the environmental costs associated with N fertilizer production. Their importance certainly increases in regions where farmers cannot afford to buy fertilizer. However, these strategies are not a guarantee of an improvement of NutUE, because N losses either by leaching and volatilization can be considerable. There is still much to learn about nutrient cycling processes in soils, especially the mobilization and immobilization turnover (MIT), and other processes that are largely microbially mediated (Goulding et al., 2008). Arbuscular mycorrhizal fungi (AMF) can impact several of these processes, including the cycling of N, its acquisition by plants or reducing N losses by leaching or volatilization (Veresoglou et al., 2012; Cavagnaro et al., 2015), as well as play other crucial roles within the agricultural ecosystem (see Box 1.3). The aim of this book is to assess the potential contribution that the intentional management of AMF can make to greater efficiency in resource use.

The natural P content of most soil in the world is small and its availability is even less due to its ready incorporation into sparingly soluble compounds of calcium and bound to Fe and Al. Therefore the need for external inputs of P is usually much greater than the uptake by the crop. The situation is aggravated in soils with a large P fixation capacity, owing to either an acid or alkaline pH. However, different species of AMF are known to be effective in establishing mycorrhiza over various ranges of pH (Abbott and Robson, 1985b). Although tropical soils are as diverse as the ones found in temperate climates (Ritcher and Babbar, 1991), many soils in the humid tropics are intensively weathered and contain iron and aluminum oxides, resulting from the desilication of clay minerals and contributing greatly to the P fixation capacity. In SSA fertilizers can cost 2–6 times more than in other parts of the world (Sanchez, 2002) and local rock phosphate can be an alternative to superphosphates, although having a much smaller solubility. Increasing SOM will also be crucial to improve P efficiency and crop yields in several parts of the world. AMF can considerably increase P acquisition by the plants, both from inorganic sources, including less soluble fractions (Cardoso and Kuyper, 2006), and from organic P (Jayachandran et al., 1992).

Water stress limits crop yields in most regions of the world and although irrigated agriculture represents only 16% of the cropland it contributes about 40% of the production (Waggoner, 1995). Water is regionally scarce, and what has been a regular pattern of global increase in the land area under irrigation, area is slowing down. One consequence is that there are important regions in the world where food production systems are already using irrigation above the natural recharge capability for the climate zone (Tilman et al., 2011). It is anticipated that climate change will exacerbate the imbalance. Agriculture now accounts for an estimated 69% of human water use. It has grown from 579 $km^3$ in 1900 to 3973 $km^3$ in 2000 and is projected to rise to 5235 $km^3$ by 2025 (Perry et al., 2009). Therefore improvement in water productivity (WP) must be a central objective in the development of sustainable production systems both for rainfed and irrigated agriculture. Improvements in WP can be achieved by a reduction in water loss as runoff, deep percolation, and direct evaporation, or by increasing the WUE of plants, that is the biomass

**BOX 1.3 Arbuscular Mycorrhiza**

Arbuscular mycorrhiza (AM) are the structural and physiological combination of a well-defined group of fungi and some 80% of land plants as well as some water plants. The identification of the fungi involved was long hampered by their obligate endotrophic habit. Now these fungi, comprised of multinucleate and largely aseptate hyphae, are grouped in the phylum Glomeromycota. One estimate suggests that probably they constitute more than 5%–10% of the soil microbial biomass (Fitter et al., 2011).

Arbuscules (Plate B1.3.1) are the intracellular tree-like branching structures that the fungi form within root cortical cells of the majority, but not all, monocotyledonous and dicotyledonous plants and gymnosperms. In some cases, the structures within the cells appear as coils of hyphae (Plate B1.3.2). Arbuscules are considered to be the essential locations for the exchange of carbon compounds from the plant for mineral nutrients, particularly phosphorus (P), from the fungus.

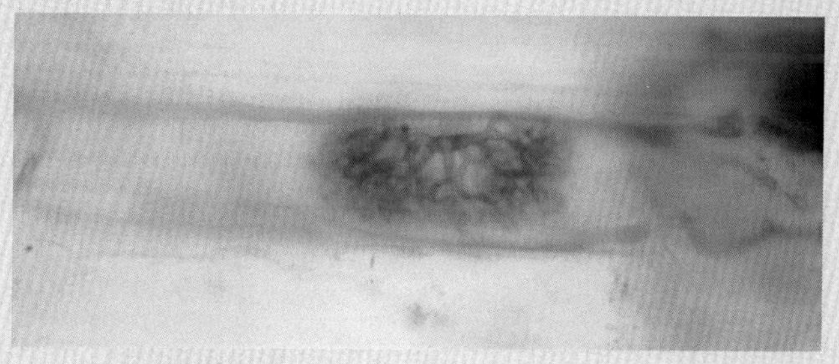

**PLATE B1.3.1** Arbuscule (400×).

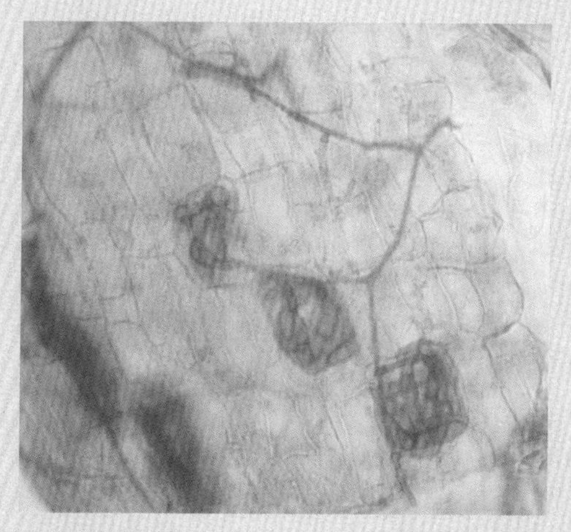

**PLATE B1.3.2** Hyphal Coils (200×).

(*Continued*)

**BOX 1.3 (Continued)**

Some of the fungi also form large oil-filled vesicles in the intercellular spaces (Plate B1.3.3). The fungi produce hyphae that explore intercellularly within the plant and those, the extraradical mycelium, that grow within the soil and can extend for several centimeters from the root surface. The latter allow large volumes of soil to be exploited for reserves of P and other nutrients, modify soil structural stability, colonize other plants, support and interact with other soil biota as well as act as a conduit for communication between linked plants (Section 3.1, Sections 5.1–5.3, Sections 6.2 and 6.3). The colonized plants can also gain from the mycorrhiza formation through the protection from abiotic and biotic stresses, both from a reduction in the effect of the stressing agent within their tissues and by the priming of their defense mechanisms (Section 3.1 and Section 5.1).

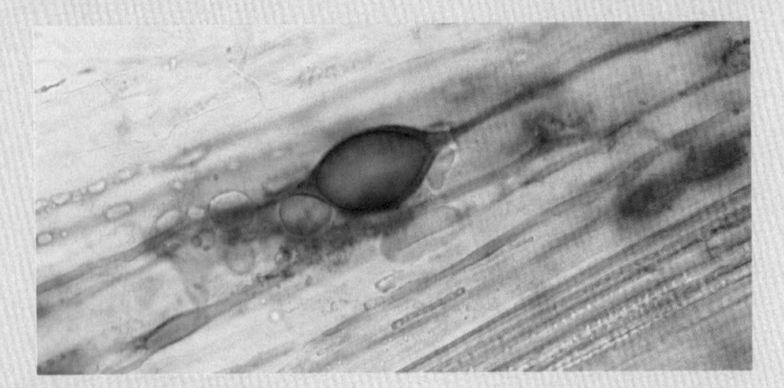

**PLATE B1.3.3** Vesicle (200×).

production per unit of water transpired. WUE depends on the plant and on the avoidance of stresses from unbalanced nutrition, disease, heat, and competition from weeds. The right choice of crop and plant breading (including GM) are important aspects to consider under sustainable intensification. The former will depend on the climate, soil, and irrigation management. The maintenance of an adequate soil structure to improve infiltration and to favor a deep rooting of the plants is an important aspect to be considered. No-till and reduced tillage systems associated with permanent soil cover by crops and residues can have positive impacts in these aspects (Shipitalo and Edwards, 1998). Increasing the SOM content improves available water-holding capacity of the soil. Water harvesting systems are also an important component for improving the soil stock under arid and semiarid conditions.

The role of AMF in improving the aggregation of soil and the resistance of the aggregates to raindrop impact and slaking is well recognized. There are several mechanisms identified, including the enmeshment of the soil particles by the hypha, the formation of proteinaceous compounds, such as glomalin, and the release of carbon into the mycorrhizosphere

(Siddiky et al., 2012) (see Section 3.1.4, Section 4.3.4, and Section 6.4.3). However, AMF are also effective in improving plant growth under conditions of water stress (Abbaspour et al., 2012).

Agronomic management and plant breeding have to include consideration of all the topics discussed earlier in developing sustainable agricultural systems for different regions of the world. In addition to the aspects of AMF described, the literature provides important information on the potential for AMF to protect plants against abiotic stress from excessive levels of Al (Rufyikiri et al., 2000), Mn (Brito et al., 2014), and salt (Navarro et al., 2014).

Biotic stresses are responsible for large economic losses due to yield constrains and the pesticides application, which is also a major concern in terms of soil and water contamination, their effects on nontarget organisms and hence on biodiversity. In a survey conducted in 13 European countries, farmers identified biotic stresses as the major concern within their production system (Ricroch et al., 2016). The pressure from biotic stresses is increasing with the intensification of agriculture systems and soil-borne pathogens are a major concern. For example it is estimated that soil-borne pathogens could be responsible for 50% of tomato losses in the United States (Lewis and Papavizas, 1991). In the case of plant parasitic nematodes, worldwide economic damage far exceeds $US 100 billion (McCarter, 2009). The use of pesticides to control soil-borne diseases is becoming increasingly restrictive as a result of environmental impacts. Alternative methods to control biotic stresses will be crucial for the development of sustainable intensification of agriculture, both to improved yields in regions where access to pesticides is not affordable, and to reduce the environmental impact of food production. Diversification of the cropping systems is playing a central role in an integrated pest management strategy. Plant breeding (including GM) will be also an indispensable tool. However, the occurrence of new pathogenic races is a continuous problem. Maize hybrids in the United States now have a useful lifetime of about 4 years, half of what it was 30 years ago (Tilman et al., 2002). Several soil microbes, including AMF, have the potential to protect plants against most of pathogens, and their intentional use as bio-protectors need to be brought into the production systems.

Despite the fact that most cultivated crop plants are mycorrhizal and all the potential benefits that AMF can confer, the intentional use of AMF within current cropping systems has been marginal. According to Whipps (2004), out of the 80 disease biocontrol products commercially available in the world, none include AMF in their composition. Several reasons are given for the lack of exploiting knowledge of AMF benefits. AMF biological diversity is very large and for all the potential beneficial aspects conferred by AMF, functional diversity exists between different AMF isolates. However, commercial inoculum of AMF is composed by a single or very limited number of isolates and cost prohibits its use in large-scale operations (Sikora et al., 2008). Under field conditions, when the stressor is already present in the soil, the role of AMF in protection is challenged by the time required to achieve an adequate level of AMF colonization (Brito et al., 2014). Moreover, native AMF are more likely to be adapted to the stress conditions present in the system and therefore more effective than introduced strains (Tchabi et al., 2010). There is a need to predict the performance of particular plant–fungus combinations under a range of environmental conditions and methods of manipulating the fungal community, by appropriate cultural practices, so as to promote the most effective isolates (The Royal Society, 2009). AMF can play an important role in sustainable intensification of agriculture by improving crop productivity, by reducing environmental impacts associated with extraction of water for irrigation, the use of fertilizers and pesticides, by protecting the soil from erosive forces and by increasing the resilience of agriculture ecosystems. If AMF are managed within the cropping system, thereby avoiding the cost of inoculation, their contribution can be equally important in both the industrial and nonindustrial regions of the world. However, in those regions, where there are several simultaneous constraints to productivity (such as high levels of P fixation, low pH with aluminum toxicity, and moisture stress) and farmers have economic challenges to pay for inputs, the relevance of AMF in improving crop production could be crucial (Cardoso and Kuyper, 2006).

## 1.3 CONCLUSIONS

The increase in global food production is an unprecedented challenge in the history of agriculture, particularly if we consider that the solutions adopted in the past, such as increasing the area of irrigated land, are only marginally available for adoption. At the same time there is an urgent need to reduce the environmental impacts of food production. It will be crucial to close the yield gap in the different regions of the world, particularly those where it is greatest due to environmental, economic, and social conditions. The world is not in a position to ignore the possible contribution from any technological solution on ideological grounds and the concept of sustainable intensification of agriculture has to be on the agenda. Among the possible solutions the intentional use of beneficial soil microbes in agricultural systems is only in its early days.

There is a much greater need than ever to find ways of exploiting the benefits from the microbes, especially AMF, in our soils and to develop tools that will help farmers implement policies related to sustainable soil use and management. Such an approach has the potential not only to increase production, while decreasing the incorporation of inputs, and to be applied to marginal soils and regions of the world where resources available for farmers are scarce.

Chapter 2

# Agronomic Opportunities to Modify Cropping Systems and Soil Conditions Considered Supportive of an Abundant, Diverse AMF Population

## Chapter Outline



It is known that there is a great functional diversity within and between different species of arbuscular mycorrhizal fungi (AMF) in terms of the benefits they may confer to host plants, such as the acquisition of nutrients or protection from biotic and abiotic stresses (Harrier and Watson, 2004; Yano and Takaki, 2005; Nogueira et al., 2007). These aspects of diversity are discussed in Chapter 4. Frequently coinoculation with different AMF isolates increases the benefits (Thygesen et al., 2004; Oliveira et al., 2006; Lax et al., 2011). Maintenance of an abundant and diverse AMF population in the soil is considered important to enhance the role that the microbes can play in managed ecosystems (van der Heijden et al., 1998b). However, there are concerns about the impact that several components of crop production systems can have on indigenous AMF populations.

*An abundant and diverse population of AMF is important to enhance their role in managed ecosystems.*

**Functional Diversity of Mycorrhiza and Sustainable Agriculture.**
**DOI: http://dx.doi.org/10.1016/B978-0-12-804244-1.00002-2**

## 2.1 COMPONENTS OF CROPPING SYSTEMS

In commercial crop production farmers need to prepare the land for seeding, protect the developing and maturing plants from pests and diseases, provide a suitable supply of nutrients and water, and ensure a timely harvest. A prerequisite for the management of arbuscular mycorrhiza (AM) in agricultural systems is the maintenance of an abundant, diverse, and mutualist AMF population in the soil. Consequently it is critical to understand how the various techniques available for use within production systems impact AMF and their diversity (e.g., Helgason et al., 1998).

### 2.1.1 Land Preparation

Land preparation involves *tillage* and, often, some additional measure of *weed control*. Tillage for sustainable production varies greatly (Table 2.1) according to local soils and climate. It has been an essential component of crop production systems as it is used to prepare the seedbed, manage soil structure, and historically has been the main mechanism used to control weeds (Hobbs et al., 2008). *Inversion tillage*, as the name suggests lifts a cut slice and turns it so that the original surface is placed at the bottom of the reconstituted layer, is both the most demanding in energy requirements and results in the most soil disruption. In part the need to use a *secondary tillage* operation, using disks or tines to create the seedbed and deal with emerging weeds, is a factor in the level of soil disruption. Although this form of land preparation is effective in disrupting existing root systems, controlling weeds and in burying straw, harvest residues or incorporating manure and other organic amendments, the mechanical disruption of aggregates and the exposure of those lifted to the surface can leave the soil susceptible to wind and water erosion.

To overcome the potential impacts of inversion tillage on soil degradation, together with the risks from compaction and rapid consolidation of the cultivated layer, numerous tillage systems have been developed, the majority since the middle of the last century, and are in use worldwide (Table 2.1). The most common feature is the absence of any soil inversion and the use of rigid or sprung tines to lift and shatter the soil. Such systems, often referred to as *reduced tillage*, may disturb the soil to the same depth as inversion tillage or only to 7–10 cm. At the furthest extreme is a system with no soil disturbance below the depth of seed placement: the *no-till system*. Under this system, it is the seeder that creates the very restricted superficial disturbance. No-till and reduced tillage systems largely rely on the use of *herbicides* to control weeds. However, the lack of disturbance greatly influences the number, type, and frequency of emergence flushes of weed seedlings, and both increased and decreased populations of weeds can result from adopting the practice (Chauhan et al., 2006; Clements et al., 1996; Mulugeta and Stoltenberg, 1997;

**TABLE 2.1** The characteristics, aims, and some essential effects of common tillage systems and potential impacts on arbuscular mycorrhiza (AM)

| Tillage Technique | Main Machines Involved | Principal Aims and Effects Achieved | Possible Effects on Mycorrhiza |
|---|---|---|---|
| Inversion tillage | Mouldboard plough, offset disks | Effects inversion of the soil. Disturbs and loosens the soil; buries weeds and residues from the previous crop, incorporates manure and organic amendments. Conventional tillage is usually associated with the use of the plough. Increased risk of soil erosion. | In the plough layer, roots, AMF mycelium, and biological channels are disrupted. Soil organic matter (SOM), less mobile nutrients and arbuscular mycorrhizal fungi (AMF) spores are distributed more uniformly. |
| Ridge till | Ridge tillage cultivator and modified planter | Soil is pushed by pairs of winged tines to form ridges up to 20 cm high. The harvest residues on the surface are removed and the top of the ridge flattened by a horizontal blade so that the seed is planted in the ridges without residue cover in the row. Contrasting soil conditions are created in the ridge and between the ridges. Residues and organic amendments are mixed with soil in the sides of the ridges each year. The ridges warm up quicker than flat land in spring, in part because the ridges drain more readily. | At the interrows there is little root development. In the ridges, SOM builds up and conditions for root development are improved. |
| Reduced tillage | Rigid or sprung tines | No soil inversion. Typically harvest residues are left on the soil surface. | The disruption of roots, AMF mycelium, and biopores is partial. Stratification of SOM, less mobile nutrients, and AMF spores are favored. |
| Deep reduced tillage | Rigid tines | Noninversion tillage to a depth greater than 100–150 mm. | The disruption of roots, AMF mycelium, and biopores are partial. |
| Shallow (reduced) tillage | Rigid or sprung tines | Noninversion tillage to a depth of less than 100 mm. | The disruption of roots, AMF mycelium, and biopores occurs in the disturbed soil. Residues remain covering the soil surface. |

(*Continued*)

**TABLE 2.1** (Continued)

| Tillage Technique | Main Machines Involved | Principal Aims and Effects Achieved | Possible Effects on Mycorrhiza |
|---|---|---|---|
| Stubble-mulch tillage | Tines or disks | Residues retained on the soil surface with soil disturbance without inversion. | Disruption extensive to the depth of the tillage tool. |
| Zero-tillage/ no-till/direct drilling | Drill with combination of tines or disks | Seed is drilled into the stubble of the previous crop with minor soil disturbance confined to the row and does not extend below seeding depth. Soil conditions are closer to the ones found in the natural environment if compaction can be avoided. At the micro-scale there is opportunity for contrasting ecological niches to develop in the soil. | Roots, AMF mycelium, and root channels are left undisturbed. Stratification of SOM, less mobile nutrients, and AMF spores is maximized. |
| Zone or Strip-tillage | Rigid or sprung tines | Zone-tillage is a development of no-tillage crop production. It consists of tillage with coulters or rigid tines (down to 15 cm below the soil surface) in a narrow strip of soil. Seeding takes place in the tilled zones, with the interrow area remaining undisturbed and retaining crop residues. | The disruption of roots, AMF mycelium, and roots channels are partial. Stratification of SOM, less mobile nutrients, and AMF spores are favored. In a large scale, contrasting soil conditions are induced between tilled and no-tilled zones. In the latter soil conditions will be similar to no-till systems. |
| Strategic tillage | Rigid or sprung tines mouldboard plough | A system predominantly using reduced tillage equipment but includes the use of mouldboard ploughing as a flexible responsive to seedbed concerns (typically weeds) or the need to incorporate organic amendments, manure, or crop residues. Avoids long-term development of soil conditions typical of natural soils. | Roots and AMF mycelium periodically disrupted. |

| | | | |
|---|---|---|---|
| Rotational ploughing | Rigid or sprung tines, mouldboard plough | Reduced tillage or no-till is used routinely for soil preparation but a plough is used at key points in the rotation. | Roots and AMF mycelium periodically disrupted. |
| Secondary tillage | Powered or unpowered tines, disks, rotary implements, rollers | Tillage practices used after the principal tool to create a seedbed and provide additional weed control. | Completes disruption of roots, AMF mycelium, and biopores. Homogenizes the distribution of SOM, immobile nutrients, and AMF spores. Enhances impacts from erosion. |
| Remedial tillage | Subsoiler tines, rotary spades, or mole plough | Tillage used when required to remove compacted layers or improve water movement. | Improves air-filled porosity and the continuity of vertical structural pores and increases the depth of rooting. |

Froud-Williams et al., 1981, 1983; Roberts and Potter, 1980; Schutte et al., 2014). This has resulted in the development of systems where inversion tillage is introduced into a program of predominantly reduced tillage (Table 2.1). A common and frequently expressed concern about no-till systems is that they require the increased use of herbicides (see, e.g., Swanton and Weise, 1991). However, with a proper weed management strategy it is possible to reduce the use of herbicides under no-till (Swanton and Murphy, 1996). If the preseeding herbicide is applied following the initial flush of weeds that germinate after the first rains, weed pressure on the following crop is considerably reduced under no-till in comparison with conventional tillage systems, such that use of post-emergence herbicides can be reduced or even avoided (Calado et al., 2010). Because the period of weed germination after the seeding of the crop is shorter under no-till systems (Calado et al., 2013), post-emergence herbicides can be applied earlier, when the weeds are younger, which allows fuller control with less of the active ingredient (Barros et al., 2008).

All the noninversion tillage systems have the potential to maintain straw and other harvest residues on the soil surface to protect it from erosion by wind or water (Towery, 1998). Systems that aim to protect the soil from water erosion by maintaining at least 30% residue cover on the soil or the equivalent of at least 1120 kg $ha^{-1}$ of flat, small-grain residues, in the case of wind erosion, are referred to as conservation tillage (Uri, 1999). Conservation no-till systems that maintain a residue cover can reduce soil losses by more than 75 times relative to those under inversion tillage (Towery, 1998). However, harvest residues can slow the warming of soils in spring (Power et al., 1986) and increase frost damage. Straw can insulate the soil from the radiant heat of the sun and restrict the convection of energy from the soil to plant leaves overnight. The intensity of tillage together with the presence or absence of straw greatly influences the physical and chemical conditions in the soil. Particularly where both primary and secondary tillage operations have been used, soil properties are more isotropic in the disturbed layer, which is formed of small aggregates. However, there can be a marked change with depth especially at and immediately below the bottom of the tilled layer. In contrast soils under no-till are typically anisotropic. Near the soil surface there can be a preponderance of horizontally aligned structural voids with a much smaller number of vertically oriented pores, often biopores and spaces between soil structural units. Typically soils under no-till are less susceptible to compaction, being more resilient to shearing forces, and the vertical pores are retained (Blanco-Canqui et al., 2010). This may also be important in maintaining the extraradical mycelium (ERM) of AMF. The biopores are frequently those created by roots and earthworms. Such channels are kept intact if the soil remains undisturbed and, when open at the surface, allow the rapid movement of water and oxygen into deeper layers as well as act as preferential routes for the growth of new roots to depth (Goss, 1991; Ehlers and Goss, 2016). The walls of the burrows can be enriched with organic matter due to the decay of dead roots and the various activities of earthworms, which also secrete organic material into their channels (Edwards et al., 1992; Stehouwer et al., 1993).

The direct action of the ploughshare, tines, or disks during inversion tillage not only fragments and mixes the soil within the cultivated layer; but it also inverts material that originally formed the soil surface on top of the layer that was not moved in the process. The material that is deposited in this way is rich in organic material coming from shed leaves, stems, stubble, or recent harvest residues, as well as living and dead fauna and the products of their activities, such as earthworm casts. With time, as this soil organic matter (SOM) decomposes, some migrates into the near subsoil (Box 2.1). In contrast with reduced tillage, particularly in no-till soil, the organic matter tends to be concentrated near the surface and then more heterogeneously distributed along earthworm and root channels as described earlier.

The disruption of larger soil aggregates during the more intense forms of tillage exposes organic matter that has been protected from turnover by soil biota. This can have the effect of increasing soil respiration because typically the organisms responsible are constrained in their activity by a lack of access to organic carbon. For example, Powlson et al. (2012) reviewed a number of reports from the United Kingdom and found that the mean difference in annual net carbon change between inversion tillage and no-till amounted

**BOX 2.1 Soil Organic Matter**

Soil organic matter (SOM) plays a crucial role in soil productivity. This applies both to regions of the world where farmers have access to nutrient inputs as well as where the availability of fertilizers is limited or climate and soil conditions prevent the efficient use of nutrients.

SOM can be partitioned into fractions that have different characteristics, turnover periods and roles in the soil proprieties. Between 10% and 40% of SOM is a more labile fraction than the rest and includes microbes. It has the greatest mineralization rate, having a half-life that ranges from a few days to a few years, and is very important in the establishment of soil physical and chemical proprieties, such as the development of soil structure and the rate of release of nutrients in forms available to the plant. Within this fraction is a physical component that can be obtained by sieving or density fractionation, called particulate organic matter (POM) with a half-life of about 50 years. POM fractions refer to coarse size fractions (>53–100 μm or 53–250 μm) or "light fractions," which have a density that typically ranges from 1.4 to 2.2 g $cm^{-3}$. It is comprised of pieces of plant material and fungal hyphae and spores, together with pollen grains. This fraction is seen to be sensitive to soil management practices (Nissen and Wander, 2003).

The more recalcitrant fraction of SOM is the results of several stages of decomposition. It is characterized by a complex of long carbon chains or rings and commonly called humus. Many microorganisms cannot use humus as an

*(Continued)*

**BOX 2.1 (Continued)**

energy source and therefore its rate of turnover is very slow. It has a half-life in soil that ranges from about 10 years to centuries. This fraction of SOM is an important contributor to the cation exchange capacity and water holding capacity of the soil. In addition, it is important in processes that determine fertilizer use efficiency. The effect of different components of the cropping system that may affect the organic matter inputs (like crop rotation and residues management) and the losses by mineralization and erosion, especially soil tillage, are mainly affecting the labile fraction of SOM.

Tillage system, crops and crop residue management are key to the content and turnover of SOM. But under no-till and other systems where crop residues are retained, the content of SOM in the soil surface layer can sometimes increase beyond the level of that explicable by the reduction in soil erosion and effects resulting from the lack of mixing. In part this may be the result of increased dry matter production because less water is lost by runoff or evaporation so more is available for transpiration by the crop. The more that plants are able to transpire freely, without stomata's closing, the greater the amount of carbon that can be fixed by photosynthesis and the larger the mass that will be returned to the soil in root material, exudates and harvest residues. The enhancement of SOM content is very efficient in improving the increase in crop production for each unit of fertilizer applied, the *fertilizer use efficiency*. For example, Carvalho and Lourenço (2014) report that under Mediterranean conditions, where climatic variability and the concentration of rainfall during the winter impaired nitrogen use efficiency (NUE), by increasing SOM from 1% to 2% NUE of applied fertilizer was doubled (Fig. B2.1.1).

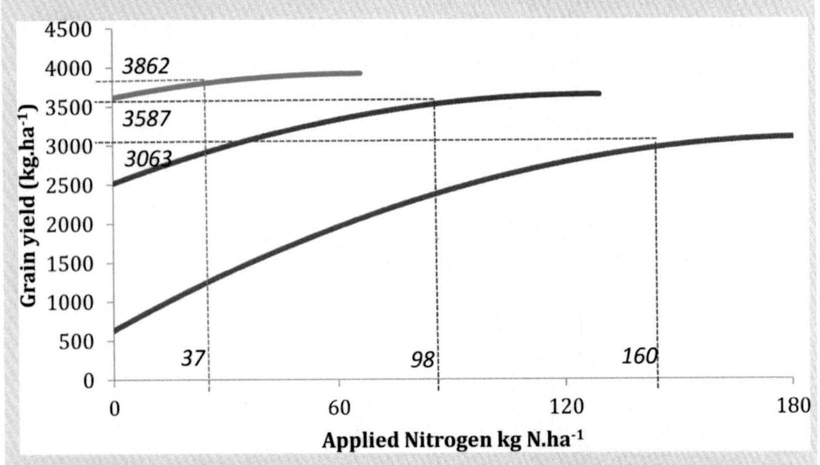


**FIGURE B2.1.1** Effect of soil organic matter on wheat response to nitrogen fertilizer application according to the fit model: Y = 631 + 35.4 N-0.07 N2 + 2718 ln(SOM)-8.6 N SOM (Y, wheat grain yield; N, nitrogen applied; SOM, soil organic matter (% 0–30 cm)). Red line, SOM 1%; brown line, SOM 2%; blue line, SOM 3%. Numbers in italics represents the economic optimum for applied nitrogen (*x* axis) and the expected grain yield (*y* axis) according to the model.

to $-315 \pm 180$ kg C ha$^{-1}$ year$^{-1}$. Six et al. (2002) used an assessment of the C associated with different soil mineral size fractions under inversion tillage and no-till in both tropical and temperate regions and obtained a value of $-325 \pm 113$ kg C ha$^{-1}$ year$^{-1}$. Assuming that the carbon input from plant material was the same in both tillage systems, the soil biota under inversion tillage annually turned over $\sim$107 g C m$^{-3}$ soil more carbon than did those under no-till. Mazzoncini et al. (2011) investigated changes of SOC in the top 30 cm of a Typic Xerofluvent loam soil profile in central Italy over 15 years. They applied three treatments: tillage, fertilizer N, and cover crops prior to spring sowing of main crops. They found that the difference between inversion tillage and no-till was -650 kg C ha$^{-1}$ year$^{-1}$. The application of fertilizer and the growing of cover crops had the effect of increasing SOC, although the impact was smaller than with tillage (Table 2.2). In the first 5 years maize was grown annually but in the next 7 years it was grown in rotation with durum wheat and then sunflower was included, with maize being grown every fourth year. Despite the changes in the components, the introduction of a rotation also increased SOC but the effect was much greater in treatments that had shown large losses of C over the first 5 years (Table 2.2). Neto et al. (2010), for the Cerrado region of Brazil, found an

**TABLE 2.2** Effects of tillage, application of N fertilizer, use of cover crops and effects of crop rotation on the changes in soil organic carbon in the top 0.3 m soil during a 15-year experiment in central Italy. In the first 5 years (A), maize was grown in each year; in the following 10 years crops were grown in rotation (B) (Based on Mazzoncini et al. (2011)).

| | Change in Organic Carbon in 0–0.3 m (g m$^{-3}$ year$^{-1}$) | | | |
|---|---|---|---|---|
| **Period** | **1993–98 A** | **1998–2008 B** | **1993–2008 A and B** | **Rotation B–A** |
| **Tillage** | | | | |
| Inversion tillage | −122.9 | 32.9 | −19.0 | 155.8 |
| No-till | 153.2 | 165.0 | 161.1 | 11.8 |
| **N-Fertilizer Addition** | | | | |
| Nil | −264.8 | 110.0 | −14.9 | 374.8 |
| Full | 130.9 | 142.0 | 138.3 | 11.1 |
| **Cover Crops** | | | | |
| No | −38.3 | 35.8 | 11.1 | 74.2 |
| Nonlegume | −42.1 | 74.8 | 35.8 | 116.9 |
| Legume-large N-fix potential | 101.2 | 137.4 | 125.3 | 36.2 |

increase of C stock in first 30 cm of the soil of 190 g C $m^{-2}$ $year^{-1}$ under an arable no-till mulch-based system, and after 12 years the SOC stock of the soil was no longer significantly different from the ones found under natural vegetation, while under a 23-year-old conventionally tilled and cropped soil the SOC stock were about 30% lower than under natural vegetation. In the Paraná state of Brazil, Neto et al. (2009) found a similar annual increment of 194 g $m^{-1}$ $year^{-1}$ by adoption of conservation agriculture based on no-till and crop rotation including cover crops.

*In comparison with inversion tillage, no-till can change soil conditions to those more conducive to AMF, including increased SOM in the surface layer, and improved water holding, but less erosion, disturbance of extraradical hyphae, and uniformity of nutrient distribution.*

In soils under no-till systems the marked stratification of organic matter also applies to less mobile nutrients, such as P. The use of no-till keeps the roots of crops and weeds intact, as it also does for the ERM network of the AMF. The stratification of nutrients and organic matter together with the lack of disruption of roots of previous crops and heterogeneity of structure provides greater opportunities for the development of contrasting ecological niches in no-till soils and hence for the survival of those AMF, such as isolates belonging to the Gigasporaceae family, less able to compete with aggressive spore producers. The conditions to some extent mimic those in natural ecosystems and potentially allow for increased AMF biological diversity and inoculum potential. In natural systems AMF taxa can be highly variable at a small space-scale, and independent of the host plant species richness. Therefore ecological niches within the soil might be an important factor in allowing different AMF to interact with the same plant root system. However, if the different fungal symbionts are spatially separated, the plant can allocate more resources to the AMF that is more beneficial to the symbiosis (Bever et al., 2009).

The positive effect of no-till in enhancing of AMF biological diversity (Jansa et al., 2002, 2003; Alguacil et al., 2008; Schnoor et al., 2011; Brito et al., 2012b), and abundance (Jasper et al., 1989b; Evans and Miller, 1990; Kabir et al., 1998) is well established in the literature. Using soil from a field under natural vegetation or after 20 years of conservation agriculture with no-till and crop rotation, Venke Filho et al. (1999) found an increase in AMF colonization rate in the agricultural soil from 23% to 66% and from 18% to 55% in maize and soybean, respectively. Another possible long-term impact of no-till on AMF abundance and effectiveness can result from the reduction in soil erosion (Habte, 1989).

Most of the studies reporting the effect of tillage systems on AMF biodiversity used spore identification and trap cultures to evaluate the AMF biodiversity. Therefore, the results may be biased in the same way as studies comparing diversity in natural and agricultural systems have been in the past, before the use of the new generation molecular tools that allow the identification of different isolates in the soil and in the roots of plants directly collected from the field (see Chapter 7). It will be important to apply these new tools in future research comparing the effect of different agricultural practices on the AMF biodiversity, if a differential effect between AMF species is to be expected, particularly a positive effect of no-till on species less easily cultured (see Section 4.4). The colonization rate of the crops under no-till starts earlier and develops faster due to the presence in the soil of an intact ERM which enhances the role of AMF on the uptake of nutrients (Martins and Read, 1997; Fairchild and Miller, 1988; Miller, 2000) and the bioprotection against excessive levels of metalloid ions (Brito et al., 2014; Alho et al., 2015). Using inoculum from a long-term no-till and conventional tillage, Köhl et al. (2014) have found that P uptake was enhanced in a mixture of grassland plants, which was consistent with the twofold greater hyphal length in the plants inoculated with AMF from no-till than conventionally cultivated soil.

The effects of intermediate tillage systems on AMF abundance and diversity are not so clear. Positive effects of reduced tillage systems have been reported (Celik et al., 2011; Säle et al., 2015) but there are also reports presenting comparative negative impacts of reduced and conventional tillage systems in relation to no-till (Kabir et al., 1998). Detailed evaluation on the effects of ridge-till systems on AMF abundance and diversity have not been found. These tillage systems have the potential to create a great diversity of ecological niches into the soil and they might be a good alternative to improve the role of AMF within agricultural systems, particularly in regions where crop performance under no-till is consistently poorer than under ploughing.

### 2.1.2 Cropping

Growing different crops in sequence on the same parcel of land appears to have been carried out from the origin of settled agriculture. The practice was known to both the ancient Greeks and the Romans (White, 1970) and its steady development occurred in Western Europe from the time of Charlemagne (AD ~800), such that crop rotation was widespread by the 11th century (Leteinturier et al., 2006).

Crop rotation has been the traditional approach to ensure that neither pests nor diseases of a particular crop build up to epidemic proportions in the soil or field environment. A crop rotation is a planned cyclical sequence of crops over time, commonly with each being grown in pure stand. A typical sequence of crops in a rotation will include monocot and dicot species, with a nitrogen-fixing leguminous plant being one of the dicots grown. Ideally the mixture of plants will include some that are shallower and others that are deeper rooting. The variation in depth of rooting contributes to maintaining good subsoil structure and both the capture and recycling of nutrients that are leaching below the root zone of shallow rooted plants. Cropping patterns other than pure stands are also grown, including growing plants in mixed stand and strip cropping (strips of two or more crops grown in the same field). There are also rotations that include crops that are not grown to maturity, such as forage, cover, or catch crops. Some mixed cropping systems include a perennial crop and the crops and weeds that grow between the individual perennial plants.

The importance of crop rotation is increased under reduced or no-till systems where the crop residues accumulate on the soil surface and are an essential element of conservation agriculture systems (FAO, 2015b). To improve the potential contribution of AMF in the agricultural systems, the design of crop rotations (including cover crops) must also take into account that the plant host is a key aspect of the maintenance of AMF biological diversity. The biological diversity of AMF depends on the ecosystem, the greatest richness is found in tropical forests while habitats with anthropogenic influence have the least (Öpik et al., 2006). Importantly the number of AMF species present seems to be related to the number of host species (Öpik et al., 2009).

*Cropping systems must recognize the significance of AMF–host relationships to improve AMF biological diversity and potential contribution to sustainable agriculture.*

Although a great abundance of AMF can be found in monocultures (Sasvár et al., 2011) the positive effect of a diversified crop rotation on AMF diversity is well established (Vestberg et al., 2005; Hijri et al., 2006). However, the inclusion of long bare-fallow (Troeh and Loynachan 2003; Johnson et al., 2003) as well as nonmycotrophic crops in a rotation (Gavito and Miller 1998; Vestberg et al., 2005; Monreal et al., 2011) can greatly decrease the AMF colonization in the following crop. The alternating of cereals and legumes seems to improve the benefits of AMF which might be related to the fact that grasses and broadleaved herbaceous plants seem to act as distinct groups of functional host plants (Lekberg et al., 2013). Negative feedback mechanisms on the AMF associated with maize and soybean (Johnson et al., 1992) seems to indicate that rotating grasses and legumes provides a good opportunity both to improve AMF biodiversity and capture more benefits of the symbiosis. Under mixed farming systems the possibility of including multi-annual leys or annual forages, consisting of closely associated plant species, is very efficient in improving AMF biodiversity in the agricultural systems (Oehl et al., 2003). Significantly, Picone (2000) found no differences in the diversity or abundance of AMF between pasture and rainforest soils from Nicaragua and Chile.

Cover crops should also be considered as a key element on the management of AMF diversity and abundance, as long as they are mycotrophic, beside other benefits generally recognized in the cropping systems. Cover crops with mixed species or rotations involving legumes, cereals, and other groups can increase the diversity as well as numbers of individual AMF. One contribution of AMF is their role in soil structure formation and stabilization. Root exudates contain the critical components for initiating symbioses between AMF and rhizobia (see Section 5.1) but also contribute to soil stability. Uptake of water by crops can also contribute to the stabilization of soil structure, especially by enhancing the efficacy of root exudates. The turnover of roots and plant litter also contribute to soil structure formation but the material with a low C:N ratio is also considered to enhance the beneficial microbes that can improve the efficacy of the symbiotic benefits from AM. Nitrogen from the breakdown of residues as well as improved soil–plant–water relations together with AMF and beneficial microbes can improve the growth of the crops (Fig. 2.1). Cover crops are particularly important to restore AMF inoculum potential in the soil after oil seed rape (non-mycotrophic), fallows or after long periods rendered unfavorable for plant growth by drought or cool temperatures (Kabir and Koide 2002; Lehman

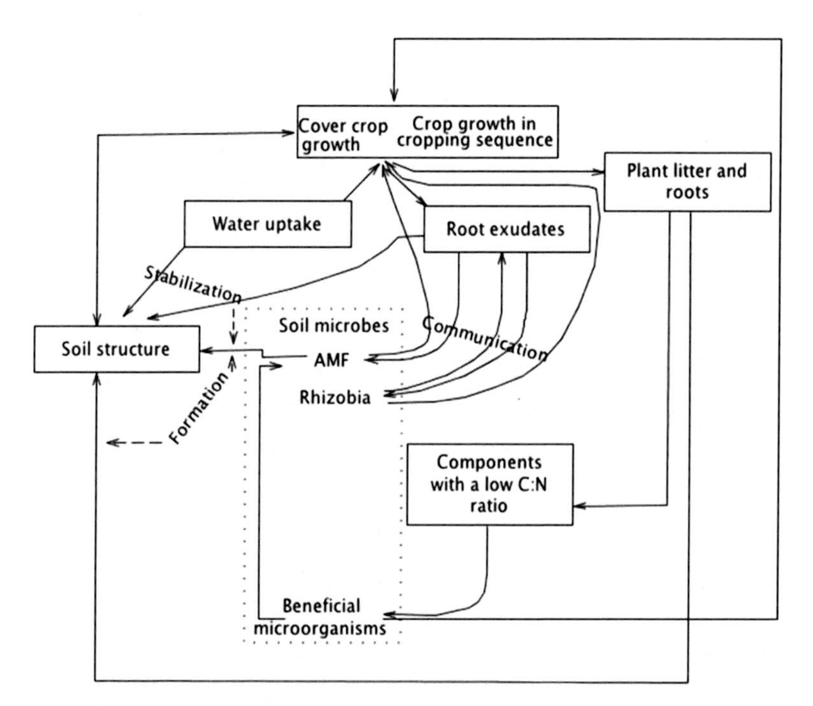


**FIGURE 2.1** Rhizosphere and mycorrhizosphere interactions under cover crops and crop rotations that encourage the presence and diversity of AMF and the benefits of these and other beneficial microbes in plant productivity.

et al., 2012). However, there is evidence that intact hyphae can survive and remain infective after a dry hot summer under Mediterranean environment (Jasper et al., 1989a; Brito et al., 2011) or a cold winter (Miller, 2000). Therefore, the ability of cover crops to enhance AMF inoculum might be increased under intensive tillage systems, although the benefits to the subsequent main crop might be even greater under no-till as the ERM developed by the cover crop remains intact and highly infective (Brito et al., 2014; Alho et al., 2015). The combination of cover crops, composed of a mixture of different plant species, and no-till might be another way of improving the AMF biodiversity within arable farming systems (Kabir et al., 1998; Kabir and Koide, 2002).

Vukicevich et al. (2016) reviewed the literature on the improvement of microbial diversity that could be applied to perennial cropping systems, such as orchards, through the use of cover crops. They considered the potential use of cover crops for enhancing the quality of the soil microbial population. In addition to the potential relationship between above and belowground diversity (van der Heijden et al., 1998a) the specific benefits associated with different plant groups were listed (Vukicevich et al., 2016). Legumes appeared to be associated with greater diversity than grass and herbs (Scheublin et al., 2004), with $C_4$ grasses being more responsive to AMF than $C_3$ (Hetrick et al., 1988). Nonmycotrophic brassicas were expected to be inhibitory of both the diversity and numbers of individual AMF, in part because of their inhibitory effects on spore germination (Schreiner and Koide, 1993). However, the use of cover crops might be limited in climatic regions where there is insufficient water to grow two crops in a year, such as Mediterranean and semiarid regions of the world. Instead of cover crops, mycotrophic weeds that germinate in the beginning of the season can increase the colonization of the crop, its growth and P uptake if they are controlled by herbicides and the crop is seeded without soil disturbance (Kabir and Koide, 2002; Brito et al., 2013b).

### 2.1.3 Application and Use of Mineral Fertilizers, Organic, and Inorganic Amendments in Crop Production

Significant priority needs to be given to the availability of fertilizer technology if we are to meet the challenge of feeding a rapidly growing world population with little opportunity to increase the area of cropped land (see Section 1.1). It is, therefore, vital that we gain a better understanding of the interactions between the use of fertilizer and the abundance and diversity of AMF within crop production systems that can be sustainable.

Plants require some 14 mineral nutrients to grow and function. They are classified according the amounts required by the plants as primary macronutrients (N, P, and K), secondary macronutrients (Ca, S, and Mg), and micronutrients (B, Cl, Mn, Fe, Zn, Cu, Mo, and Ni). The different nutrients should be in balance to achieve maximum growth, according to Liebig's law of the minimum. Whenever the nutrients reserve in the soil are not enough to meet crop demand external input must be applied to achieve the potential yield, that is, the yield that can be obtained according to the genetic potential of the crop and the other environmental factors, like temperature, radiation, and water. Because N, P, and K are the most frequent limiting nutrients for crop production they are the ones most frequently applied, especially N and P. At the same time these two nutrients are also the ones with a greater environmental impact (nitrates and phosphates in relation to water contamination and nitrous oxides in relation to atmospheric pollution and climate change). To maintain or improve crop yields without increasing environmental concerns (see Box 1.2), the current focus is aimed at balancing the crop response to fertilizers with nutrient budgets at the field scale – a strategy known as the rational use of fertilizers. In this approach nutrient use efficiency (NutUE), which can be defined as the proportion of applied nutrients adsorbed by the crop, has become a critical aspect of sustainable agriculture (Goulding et al., 2008). The nutrient reserves of the soil comprise the mineral and the organic phases, the latter being particularly relevant for N, P, and S. When soil fertility limits growth, nutrients must be applied as inorganic fertilizers or organic amendments to improve crop yields and, because NutUE of a crop is always less than 1, a small positive balance is inevitable and usually considered to be desirable. On the other hand if the level of a nutrient in the soil exceeds crop requirements to guarantee the potential yield, then the nutrient is likely to be lost to the environment, even in the absence of any external inputs. The ideal soil nutrient level occurs when the mass of the nutrient applied equals the amount removed by the harvest crop and is sufficient to ensure the desired yield (Janssen and Willigen, 2006). Traditionally organic amendments were mainly animal manures but now there is a very long list of materials, including municipal biosolids, garden waste, and untreated or composted materials of plant and animal origin coming from the food supply and processing industries. A newer material receiving considerable attention is biochar, a biomass-derived black carbon produced by pyrolysis of organic material, which occurs naturally as the result of vegetation fires. This carbon is very recalcitrant to microbial decomposition and can contribute to the enhancement of soil fertility and water holding capacity, although positive effects on crop performance can be variable (Gaskin et al., 2010). The effects of biochar on AMF abundance can be either positive or negative. Furthermore, there is insufficient information on its effects on mycorrhizal fungal species composition (Warnock et al., 2007).

It is commonly considered that the application of both P (Kahiluoto et al., 2009; Wang et al., 2009) and N (Egerton-Warburton and Hallen, 2000; Wang et al., 2009) fertilizers reduces the colonization of crops by AMF. Furthermore, the reduced level of colonization may also result in a decline in the abundance AMF spores, species richness, and the range of mutualism that a host plant can access in support of its acquisition of nutrients, such as P (Johnson, 1993), although the effect also depends on the AMF species (see Section 4.4). Tawaraya et al. (1996) demonstrated that variation in soil P had no effect on the germination of spores of AMF or the growth of hyphae. There are numerous reports (see Braunberger et al., 1991) showing that as the concentration of P in soil increases, the number of hyphopodia, also known as appressoria (the structures created by the fungi at points of entry into the host root that adhere to the root surface and allows the penetration of the cell wall – see Section 5.1.1, Box 5.1) decreases. However, the cause of the reduction is not certain; whether it is the concentration in the soil or the plant and, if in the plant whether it is the level in the shoot or the root. Furthermore the level of P supply also affects the rate of exploration of a root by the intraradical fungal hyphae (Braunberger et al., 1991). One significant factor is the effect of the improved phosphorus supply on the rate of growth of the roots (see Box 2.2) relative to that of the intraradical hyphae.

Although negative effects of P on AM formation may be dependent on plant P content (Braunberger et al., 1991), the reduction of AC in maize seems to be more pronounced in the zone where P is placed, with differences being related to the concentration of P in different parts of the root system (Lu et al., 1994). Therefore, banding of P fertilizer offers a strategy to reduce the negative impacts (Miller et al., 1995a). However, under no-till systems for small-grain cereals, broadcasting of P fertilizer can have a smaller impact on AC than the deep banding (Barbieri et al., 2014). This result could originate from the smaller plant P content at tillering that did not affect wheat yield but was associated with an enhanced AC of the crop at earlier growth stages. A possible explanation for the differences in the results might be the effect of tillage system on the distribution of P in the soil. Inversion tillage systems tend to result in the P content of the soil being fairly uniform within the plough layer, and therefore the banding application of the fertilizer is an opportunity to create localized differences in its concentration in the soil. Under intensive agricultural systems, deeper soil layers with smaller P concentrations could encourage a more diverse AMF community than the superficial soil layers (Gai et al., 2015). Under no-till systems the strong P stratification results in smaller concentrations occurring at shallower depths than with inversion tillage, with broadcasting of fertilizer enhancing this effect.

Reports on the impact of mineral N addition on root colonization by AMF are also contradictory (see Sylvia and Neal, 1990). There are indications of differences between additions of $NH_4^+$-N and $NO_3^-$-N on colonization but Thomson et al. (1986) found that pH changes were the major cause. The availability of mineral N and P in the soil can have a marked effect on root colonization. In a series of three experiments Sylvia and Neal (1990) found that when onion (*Allium cepa* L.) roots were deficient in N, colonization was unaffected by variation in soil P but addition of N reduced the

**BOX 2.2 Arbuscular Mycorrhiza (AM) Colonization**

The relationship between P (or other nutrients) availability in the soil and AM colonization is not straightforward. Colonization is usually determined as the percentage of the root colonized by arbuscules (AC), hypha, or vesicles and typically expressed as a rate. However, the concentration of P available in the soil affects root length. Therefore whenever P (or other nutrient) application promotes the growth of the plant and consequently root length, a reduction in the value of colonization rate results even though there may be no impact on the ability of the AM fungus to infect the roots. However, if colonization is evaluated according to the colonized root density (CRD), which is the product of the percentage of roots colonized and the root length density (the length of root per unit volume of soil) the apparent impact of P can disappear (Table B2.2.1).

Differences in root growth and development occur between species under the same conditions and in the presence of the same assemblage of AMF. Using CRD as the measure of colonization, Carvalho et al. (2015) showed that even apparent differences between plant species in the level of colonization are no longer evident. The application of P fertilizer to give a concentration of 17.3 mg P $kg^{-1}$ in the soil, significantly ($P<.001$) reduced AC in maize from 55% without P addition to 34% at 21 days after planting, but increased root density from 2.8 to 7.2 cm $cm^{-3}$ of soil. However, CRD was not affected (Table B2.2.1).

**TABLE B2.2.1** Effect of P applied to the soil (mg P $kg^{-1}$ of soil): on the arbuscular colonization (AC, % of root colonized by arbuscules), on the root density (RD, cm of root $cm^{-3}$ of soil), and on colonized root density (CRD, cm of root colonized with arbuscules $cm^{-3}$ of soil) of maize (Values followed by the same letter are not significantly different at $P = 0.05$. Adapted from Brito et al. (2013a)).

| P Applied | AC | RD | CRD |
|---|---|---|---|
| 0 | 55.3 a | 2.8 c | 2.1 |
| 5.8 | 48.4 ab | 4.7 b | 2.7 |
| 11.5 | 42.2 b | 6.2 a | 3.0 |
| 17.3 | 33.9 c | 7.2 a | 2.5 |

proportion of roots colonized. There is also evidence that the effect of N and P fertilizers can depend on the particular AMF species present (Bhadalung et al., 2005).

*There is evidence that negative effects of N and P fertilizers on AMF colonization can be interactive and depend on the species present.*

There are, however, results in the literature that contradict the view that fertilizer application negatively impacts AMF. For example, there is evidence of rates of fertilizer application that are compatible with the maintenance of heavy crop yields but do not affect AMF abundance or diversity (Hammel et al., 1994; Khalil et al., 1992; Thingstrup et al., 1998; Tian et al., 2013b; Martinez and Johnson, 2010; Porras-Alfaro et al., 2007). Jansa et al. (2014) did not find evidence that available P levels impacted AMF community profiles in their investigation of 154 agricultural soils. The observed effects on AC associated with the interaction between P and tillage cannot be related simply to nutrient stratification. When AMF colonization starts from an intact mycelium the P acquisition of the crop is enhanced, even when high levels of P are available in the soil, and despite its effect on the colonization rate (Miller et al., 1995a).

One important consideration is the quantity of nutrients to be applied. Several studies indicate that the negative effects of nutrients only occur at the highest levels of application (Miranda and Harris, 1994; see Table B2.2.1). In contrast there are reports that long-term nutrient depletion can have negative effects on AMF abundance and even the loss of ability of AMF to establish true mutualistic partnerships (Antunes et al., 2012). Field evaluation of the long-term effects of the rational application of different fertilizers is required to clarify how they impact on AMF abundance, biodiversity, and functional diversity. Secondly it has to be understood that most of the studies comparing the effects of large nutrient inputs were carried out in the presence of factors, such as intensive tillage and crop monoculture, which are recognized as being deleterious to AMF. For example, one study in central Europe by Oehl et al. (2003) compared the impact of land use intensity on AMF diversity. The number of AMF species present in soil was considerably larger, significantly so, from land under a 7-year crop rotation compared with that from land under a maize monoculture, when the same level of mineral P fertilizer was applied. Baltruschat and Dehne (1989) found that even when high levels of fertilizer N were applied AMF inoculum potential was enhanced by crop rotation and green manure. Toljander et al. (2008) found an increase in the number of AMF isolates in the roots of maize following a green manure crop and Douds et al. (1997) found significant positive effects of the crop rotation on AMF abundance even when large amount of nutrients where applied. Studies comparing the effect of tillage system on AMF showed that, even when liberal amounts of P were applied, no-till still had a significant effect on hyphal density (Kabir et al., 1998; Köhl et al., 2014). Furthermore, Jansa et al. (2002), Sheng et al. (2013), and Köhl et al. (2014) reported that the diversity and structure of AMF communities with species of non-*Glomus* AMF were more abundant in the no-till soil. In a study comparing no-till or shallow rotovation and with a large supply of fertilizer P, seeding of the highly mycotrophic black medic (*Medicago lupulina* L.), sown as a cover crop, significantly increased colonization by AMF of leek (*Allium porrum* L.), which was the main crop (Sorensen et al., 2005). Lehman et al. (2012) also found beneficial effects of cover crops in increasing AMF inoculum potential under field conditions, particularly where the soil had a large N and P content. Porter et al. (1978) found that the rate of superphosphate fertilizer applied to a long-term pasture in Australia, which was comprised of a diverse plant community and received no soil disturbance, had no impact on the abundance of AMF species or their ability to colonize and form effective mycorrhiza. Similar results are presented by Shi et al. (2016) after 4 years of annual application of 40 kg P $ha^{-1}$ to timothy (*Phleum pratense* L.) swards in Canada.

There are numerous reports of the benefits of organic agriculture on AMF abundance and diversity and frequently the effects are attributed to the use of organic P (Aryal et al., 2007; Lee and Eom 2009; Gosling et al., 2010; Oehl et al., 2003; Bedini et al., 2013; Säle et al., 2015), although the benefits might depend on the type of organic amendment (Douds et al., 1997) and the amount applied (Cavagnaro, 2014). However, in most of these studies several differences in the cropping systems described make it difficult to interpret the results. Ryan and Ash (1999) evaluating the effect of mineral and organic P application to the soil from a long-term pasture in Australia, where all the other components of the cropping system were the same, concluded that after 17 years there was no indication that the biodynamic and conventional soils had developed substantially different processes to enhance plant nutrient uptake or that the indigenous AMF differed in their tolerance to applications of soluble nutrients, although extractable P was 2–3 times greater where P was supplied by conventional fertilizer. Loss of taxa under intensive conventional agriculture rather than changes in species composition appears to explain apparent benefits from organic agriculture on AMF abundance and diversity (Verbruggen et al., 2012). However, Säle et al. (2015) only found a greater number of AMF species under organic farming when reduced tillage was used but under inversion tillage no differences in the number of AMF species were found between organic and conventional systems. Bilalis and Karamanos (2010) found that no-till increased AMF colonization rate on organic grown maize in relation to conventional tillage and Eason et al. (1999) found that, although a greater AMF diversity was found in cultivated soil under organic farming, some highly effective AMF isolates were found in high-input systems. Alguacil et al. (2014) found that AMF diversity in the roots of ornamental peach (*Prunus persica* L.) were greater in the treatment combining organic amendments and inorganic fertilizer with chemical pest control. Whatever the reasons behind the benefits of organic farming on AMF abundance and diversity, the accumulation of nutrients within SOM is certainly important from the perspective of sustainable agriculture (Carvalho and Lourenço, 2014) and have a beneficial effect on the AMF colonization (Quintero-Ramos et al., 1993; Wang et al., 2013). The contribution of AMF to plant P uptake and growth seems to be greater when the nutrients are organically supplied rather than by mineral fertilizers even at the same plant P content (Kahiluoto et al., 2012). Therefore, the large-scale adoption of strategies to improve SOM, like the adoption of reduced or no-till systems, the maintenance of crop residues on soil surface, the inclusion of cover crops and the use of organic fertilizers, would certainly make a major contribution to improve the sustainability of agricultural systems. Such strategies would also enhance the potential role that AMF can play, either in agriculture systems where excessive levels of fertilizer application are a concern for water quality and AMF biodiversity, or in those regions of the world where inputs of nutrients are only moderate or even too small and consequently cause concern for soil productivity.

### 2.1.4 The Application of Pesticides

Evaluation of how the long-term use of pesticides alters AMF abundance and diversity is difficult, with published results comparing short-term effects or confounding the use of pesticides with several other elements of a cropping system. Axenic experiments are of little interest in studying the effect of pesticides on AMF under field conditions, as it is difficult to separate their effect on each partner of the symbiosis. Fungicides are the pesticides that potentially are the most deleterious to AMF. However, there is no clear trend from their use, especially when applied at field rates, and their systemic translocation within plants does not appear to be related to any effects on mycorrhiza formation (Trappe et al., 1984). Fungicides applied as seed coating are probably more detrimental to AM development than fungicides that are applied when plants are already mycorrhizal (Plenchette et al., 2005). Repeated use of soil fumigants can have a negative effect on AMF and can result in a complete absence of colonization (Cavagnaro and Martin, 2011), although AMF species richness and diversity can recover at the end of the season after a single application (An et al., 1993). The effect of herbicides on AMF colonization is also not consistent and might depend on the tillage system, with the effect being negative only under conventional tillage (Santos et al., 2006). Negative effects of two herbicides on AMF colonization were reported when the herbicides were directly incorporated into the soil (Abd-Allaa et al., 2000), but the absence of effect has also been reported (Ryan et al., 1994; Brito et al., 2013b).

The importance of conservation and no till agriculture to the development of sustainable intensification (see Box 1.2 and Box 2.1) and for the promotion of AMF abundance and diversity (see Section 2.2), make any possible concerns about effects of glyphosate on the infectivity of AMF worthy of particular attention. Negative effects of glyphosate application on the host plant and its colonization with AMF have been reported (Zaller et al., 2014). However, within agricultural systems glyphosate is applied to control weeds or cover crops before seeding, so it is the effect on AMF in the different elements of the cropping system, particularly the next crop, that is really important to evaluate. Brito et al. (2013b) studied the infectivity of the ERM associated with target weeds following the application of glyphosate (a systemic herbicide) or paraquat (contact herbicide). A negative control was included, where the weeds were killed by soil disturbance rather than herbicides. The AMF colonization rate in the following crop was similar for the herbicide treatments with the contrasting modes of action but the values were significantly greater than that for the disturbed soil treatment, which received no herbicide. Brito et al. (2014) and Alho et al. (2015) have also reported that the colonization rate of the second plant in a succession was always significantly enhanced when the ERM associated with the first plant was kept intact, even though glyphosate was used to control the first plant. It is also significant that in the work of Brígido et al. (2017), there was no effect of glyphosate on AMF diversity. Concerns have also been expressed about possible negative impacts of glyphosate on spore viability; both when active ingredient is applied to the plant foliage (Druille et al., 2015) or directly to the soil (Druille et al., 2013). However, Brito et al. (2014) and Alho et al. (2015) found no evidences of negative effects of glyphosate application, either directly to the soil or to the foliage of plants, on spore viability or infectivity of colonized root fragments.

*Reducing biocides is a central feature of sustainable intensification.*

From the available information it is reasonable to assume that the impact of pesticides on AMF is smaller than the one resulting from the use of fertilizers. However, another important aspect related to the impact of pesticides on AMF, is that reduction in biocide use is a central feature of the sustainable intensification of agriculture. A diversified crop rotation is considered a key element in this approach, in terms of both disease (de Boer et al., 1993) and weed control (Beckie, 2006). The use of genetic tolerance or resistance together with the modification of other aspects of the production system, such as seeding time, have to be included in integrated pest management strategies with sustainable agriculture to limit the amount of pesticides applied.

## 2.2 KEY ASPECTS OF AGRICULTURAL SYSTEMS ON DIVERSITY OF MYCORRHIZA

Different elements of the cropping system can negatively impact the abundance, diversity and mutualism of AMF, especially when several such components are present simultaneously, such as intensive tillage, large applications of fertilizer and monoculture cropping. However, there are also opportunities to develop strategies that can minimize the negative effects on AMF while maintaining heavy yields. Oehl et al. (2003) compared the impact of seven different agricultural systems: three low-input species-rich grasslands, two low-input arable cropping systems (one organic other conventional), and two high-input maize monoculture systems in three different European countries. The two low-input arable systems were based on a 7-year crop rotation but with some differences. The rotation in the organic system included 2.5 years of a permanent grass–clover ley, whereas in the conventionally managed system only annual crops were grown, including the nonmycotrophic rapeseed (*Brassica napus* L.) crop, which was not grown in the organic system. In the organically managed arable system, AMF diversity was similar to the one found in permanent grassland, which is surprising and shows that AMF diversity and arable farming can be compatible. The conventional low-input arable system showed a slight but significant reduction in AMF diversity, as expressed by the Shannon–Weaver diversity index, compared with the organic system but diversity was significantly greater than in the intensive maize monoculture system, for identical levels of available P in the soil. In the work of Oehl et al. (2003) AMF identification was based on spore morphology, either from field samples or trap culture. However, when the AMF diversity of these two fields was assessed by variable regions of the ribosomal RNA genes, extracted from the roots of wheat and maize growing in the field, no significant differences were detected between the conventional and organic low-input systems (Hijri et al., 2006). The inclusion of permanent leys in the crop rotation is certainly a unique opportunity to support AMF diversity. Although rainforest is recognized as being the most diverse ecosystem on Earth, its conversion to pasture in Nicaragua and Chile does not necessarily impair AMF diversity and abundance in the soil (Picone, 2000).

> *AMF diversity under arable farming can be similar to that of permanent grassland if the right practices are adopted.*

In addition to the close association of different plant species, permanent leys allow a long period when soil tillage is absent, which is also supportive of AMF diversity even with different levels of available P in the soil resulting from conventionally and organically managed permanent pastures (Ryan and Ash, 1999). Under mixed farming systems it is obviously a sensible recommendation to include permanent ley of consociated grasses and legumes in the crop rotation. Consociations of grasses and legumes can also be used for forage production. In this case, if no-till is adopted together with an integrated nutrient management plan, where both organic and inorganic sources of nutrients are considered but only applied at rates that avoid excessive levels of P in the soil, there are reasons to believe that the AMF species richness, abundance and even mutualism can be maintained in levels identical to natural ecosystems. In a study carried out in Argentina the response of both maize and tomato to inoculation with material obtained from agricultural soils was greater, both in terms of colonization rate and growth enhancement, than when the natural ecosystem was the source of inoculum (Islas et al., 2014). Under arable farming systems consociations of grasses and legumes could also be used as cover crops to replace the role of permanent leys. Moreover, if appropriate tillage techniques are adopted to maintain the integrity of the ERM (see Section 2.1.1) associated with the cover crop, the colonization of the next crop starts earlier and develops faster, thereby enhancing the role of AMF on crop nutrient uptake or its protection against stresses. In regions where there is enough time, from the beginning of the rainy season and the sowing of the crop, for the weeds to germinate and became well colonized, the natural vegetation can also be used to support the development and mutualism of the symbiosis.

The evidence suggests that it is not the use but the *excessive* application of nutrients, both in mineral and organic form, which negatively impacts AMF diversity. The growing of cover crops, the maintenance of crop residues on the field, the use of organic amendments in replacement of mineral fertilizers and the adoption of no-till can greatly improve SOM and provides a strategy for reducing the application of fertilizers, thereby allowing further improvement in AMF diversity, abundance and mutualism. The anisotropic distribution of P in the soil seems to favor AMF colonization of the root systems in the impoverished zones of the soil and, therefore, fertilizer distribution needs to vary according to the tillage system. Another possibility for increasing the differences in soil P content within a field is the use of twin-line planting technology, where the fertilizer is only applied between the two crop rows.

The recovery of AMF diversity in intensively managed systems seems to be possible without the need for inoculation. For example, An et al. (1993) found that an AMF population was able to recover after only one season following fumigated with 67% methyl bromide/33% chloropicrin. Oehl et al. (2005) identified that AMF diversity was large below plough depth and differed from the one found in top soil of intensively managed maize crops, and this reserve of AMF biodiversity in deeper layers might support the recovery of AMF richness and diversity after conversion of intensive agricultural systems. Säles et al. (2015) also found that AMF spore density and species richness were very similar at a depth of 30–40 cm in soils under different agricultural systems.

*It is not the use but the excessive use of fertilizers that negatively impacts AMF diversity.*

## 2.3 CONCLUSIONS

The evidence indicates that the rational use of applied nutrients (both organic and inorganic), which is needed to maintain soil productivity, is compatible with maintaining an abundant and diverse AMF population if other elements of the cropping systems, such as the use of reduced or no-till and a diversified cropping sequence, are also supportive. Increasing SOM plays an important role to reduce fertilizers application, crop diversification will help to reduce pesticides inputs and smart strategies can reduce herbicides use under no-till. Under such systems if no-till is adopted for all the crops and only moderate amounts of fertilizers are applied it might be possible to improve AMF diversity to levels identical to the natural ecosystems. Although rainforest is the most diverse ecosystem, its conversion to pasture does not have to reduce the AMF diversity and abundance in the soil. In arable farming systems cover crops, consisting of several compatible plant species, it might be possible to replace the role of diverse leys and forage crops in mixed farming systems. There is an urgent need to use the new generation of molecular tools for the evaluation of effects of cropping systems on biodiversity of AMF under field conditions, especially on AMF from different functional groups.

Chapter 3

# The Roles of Arbuscular Mycorrhiza and Current Constraints to Their Intentional Use in Agriculture

## Chapter Outline



Symbiotic arbuscular mycorrhiza (AM) likely made possible the conquest of land by the first bryophyte-like plants ~470 million years ago (Selosse et al., 2015) and now colonize more than 80% of plants, including angiosperms, gymnosperms, pteridophytes, bryophytes (Wang and Qiu, 2006), and also aquatic plants (Søndergaard and Lægaard, 1977). Arbuscular mycorrhizal fungi (AMF), the microbial symbionts, are present across all soil types and biomes (Brundrett, 2009), in natural and anthropogenic ecosystems. These features are, according to present knowledge, unique among other mutualistic symbioses and account for the unquestionable importance of AM in earth's ecosystem.

*Significant benefits to the host plants are derived from the activities of the extra-radical mycelium.*

## 3.1 BENEFITS OF ARBUSCULAR MYCORRHIZA

Many benefits that accrue to plants from their association with AMF are directly related to the increased volume of soil that can be explored by AM

Functional Diversity of Mycorrhiza and Sustainable Agriculture.
DOI: http://dx.doi.org/10.1016/B978-0-12-804244-1.00003-4

plants by means of the extraradical mycelium (ERM), which takes up the nutrients from outside the rhizosphere within the bulk soil. The nutrients are then transported via the hyphal network to host plant root, where they are passed to the plant in exchange for carbon. AMF hyphae are longer and thinner than root hairs (RHs) (Plate 3.1), commonly reaching greater distances from the root and getting into soil pores inaccessible to RHs.

Sieverding (1991) estimated that for each centimeter of colonized root there is an increase of 15 $cm^3$ in the volume of soil explored, this value might increase to 200 $cm^3$ depending on environmental conditions. George et al. (1995) considered that the ratio of the length of AM fungal hyphae to that of roots in soil could be 100:1 or even greater.

The enhanced volume of soil explored, together with the ability of the ERM to absorb and translocate nutrients to the plant, results in one of the most obvious and widely recognized advantages of mycorrhizal formation; the ability to take up more nutrients, especially those that have limited mobility in soil, such as phosphorus (P). In addition, the extraradical hyphae stabilize soil aggregates by both enmeshing soil particles (Rillig and Mummey, 2006) and as a result of the production of substances that link soil particles together (Goss and Kay, 2005), leading to a greater stability of soil structure.

In addition to nutrient acquisition, many benefits are associated with AM plants; alleviation of water stress (Augé, 2001), increased efficacy of N-fixation by legumes (Azcón and Barea, 2010), protection from root pathogens (Harrier and Watson, 2004; Whipps, 2004), tolerance to toxic heavy metals (Heggo et al., 1990; Ahmed et al., 2006), tolerance to adverse

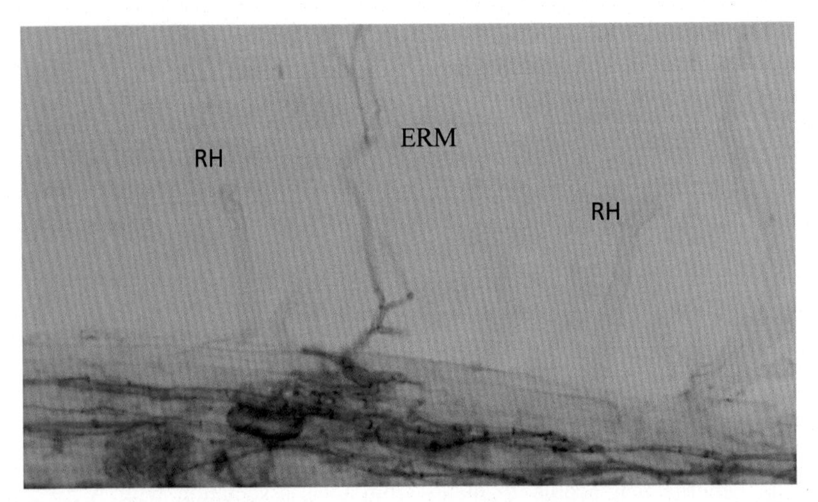


**PLATE 3.1** AMF colonized root (200 × ). Root hair (RH) and extraradical mycelium (ERM). ERM is longer and thinner than a RH and is branched.

temperature, salinity, and pH (Sannazzaro et al., 2006; Yano and Takaki, 2005; Campanelli et al., 2013), and better performance following transplantation shock (Meddad-Hamza et al., 2010). Not all the beneficial effects observed in AM plants are related to the greater soil volume explored through the ERM. Physiological and morphological changes can also occur in the host plants colonized by AMF, especially, but not solely when they are growing in the presence of stresses. The production of plant growth regulators (plant hormones), including indole-3-acetic acid (IAA), indole-3-butyric acid (IBA), and jasmonic acid (Fernández et al., 2014; Cameron et al., 2013), and several other plant hormones (Foo et al., 2013), is modified, usually resulting in a more favorable condition and greater homeostasis, for the host plant. Plant genetic expression can also be altered by AMF colonization, e.g., the mechanism governing the expression of inorganic P ($P_i$) transporters is changed, even after a very weak colonization by AMF (Harrison, 1999a; Poulsen et al., 2005).

A survey by McGonigle (1988) of 78 published field investigations concluded that increased AMF colonization resulted in an average yield increase of 37% and a similar analysis of 290 published field and greenhouse studies, carried out by Lekberg and Koide (2005), determined that increased colonization resulted in a 23% enhancement of yield. These are not trivial impacts and should be considered in the framework of sustainable intensification of cropping systems (see Section 2.1.2).

### 3.1.1 Acquisition of Mineral Nutrients

Phosphorus is one of the main nutrients for plant growth and although frequently present in the soil it is not readily available for plants mainly due to the formation of sparingly soluble compounds with cations (Fe, Al, and Ca depending on pH) or adsorbed on soil surfaces (Holford, 1997) (see Section 1.2). In addition, because its transport through soil is slow (mainly by diffusion), an area of P depletion is easily created around a plant root that limits the rate of P absorption. However, in AM plants the size of the soil P depletion zone is significantly larger. For onion plants, Rhodes and Gerdemann (1975) found that *Glomus fasciculatum* extended the uptake zone for phosphate to at least 70 mm from the root surface and Li et al. (1991) found differences in concentration that extended more than 10 mm in some cases but in others this difference was greater than 110 mm, depending on the plant host colonized and the fungal isolate involved. The same authors estimated that 80% of plant P could be supplied by AMF hyphae from as far as 100 mm distance when root exploration had been restricted. After a complex set of physiological, cell, and molecular biology analyses and also taking account of the genetics of P uptake, Bucher (2007) concluded that soil P availability together with the formation of P depletion zones around roots and their mycorrhizas are the major physical parameters determining plant

P acquisition efficiency. Additionally, phosphate ions diffuse faster into hyphae than into RHs, due to the larger affinity for P uptake and the smaller threshold concentration for absorption into the hyphae (Bolan 1991).

Although plants colonized by AMF have been shown to respond positively to the application of insoluble forms of $P_i$, such as rock phosphates, it is not unequivocally established that AMF hyphae absorb forms of P other than soluble anions (phosphate ions: $H_2PO_4^-$ and $HPO_4^{2-}$). AMF hyphae may help to increase the inorganic P availability (solubilization) as a result of the biochemical (specifically pH) changes they promote in the mycorrhizosphere environment (Tawaraya et al., 2006) or due to the microbial population accompanying AMF hyphae or spores (Jayachandran et al., 1989; Toljander et al., 2006; Antunes et al., 2007; Frey-Klett et al., 2007; Alonso et al., 2008; Bender and van der Heijden, 2015; see Section 5.1.2). Like many other soil microbes, AMF can also be involved in the very slow process of mineralization of organic P forms, through the exudation of phosphatase enzymes (Koid and Kabir, 2000).

Reports of enhanced P acquisition of AM plants relative to non-AM plants are numerous, and the extent of the difference is often stated to be dependent on the host plant, the fungal species, or isolates (Jakobsen et al., 1992b; Kahiluoto, 2004; Smith et al., 2004; Jansa et al., 2005; Pandey et al., 2005; Ehinger et al., 2009). Length-specific P uptake by hyphae seems to be rather constant within each AMF species, and the functional diversity found is associated with the fungus growth pattern and its spatial distribution (Drew et al., 2003; Munkvold et al., 2004). These authors stress that the large intraspecific diversity observed for mycelium growth and improvement of P uptake means that AMF communities of low species diversity may still contain considerable functional heterogeneity. Also the source of AMF propagules can influence P acquisition: Fairchild and Miller (1988) and Goss and de Varennes (2002) reported greater P accumulation, following AM colonization starting preferentially from ERM compared with other types of propagule (see Section 6.1).

*The role of AMF in enhancing the P supply to plants is of major importance but phosphatic fertilizer application has much less effect on the colonization of host plants than does propagule type or host characteristics. Earlier conclusions that AMF diversity is reduced by P addition might need to be revised.*

A widespread idea is that accumulating P in soil can lead to changes in AM fungal species composition and decreased the size of the AM fungal communities (Kahiluoto et al., 2001; Kahiluoto, 2004). Additionally, fertilizer P application seems to select for inferior AMF mutualists (Johnson, 1993) and, as a consequence, AM colonization decreases and may either be eliminated or lead to growth depressions (Amijee et al., 1989; Son and Smith, 1988; Kahiluoto et al., 2000; Harrier and Watson, 2003). However, according to a metaanalysis of context dependency in the response of plants to inoculation with AMF (Hoeksema et al., 2010), P fertilizer application is of relatively less importance than other variables, such as the inoculum complexity or the plant functional group (see Box 3.1).

Under field conditions application of P fertilizer is frequently made in bands close to the seed row, thereby providing adequate nutrition during the early growth stages of the new crop. Although colonization of roots growing in this zone may be markedly reduced, AM colonization of roots growing away from the band may not be reduced to the same extent. This latter portion of the root system may be an important contributor to P nutrition at later growth stages (Miller et al., 1995a). In fact the relationship between the provision of P and AMF presence is not always described as being obvious and some authors have shown that, despite a high level of P application, the host plant can still indicate benefits from mycorrhiza formation. For example, Schweiger and Jakobsen (1999) demonstrated a considerable contribution of indigenous AMF to overall P uptake of field-grown winter wheat, even at

**BOX 3.1 Inoculation With Arbuscular Mycorrhizal Fungi (AMF)**

**Experimental-Scale Use of Inoculum**

In practical terms our understanding of the benefits accruing from mycorrhiza is largely dependent on comparative studies, made between mycorrhizal and nonmycorrhizal plants or between different levels of AMF present in the host plant. When working with soil as the growth medium, the reference or control treatments require the reduction or elimination of the indigenous AMF population, which can be achieved by several methods that need to be appropriate to the investigation (Brito et al., 2009). However, many of the test treatments have to rely on inoculation with AMF. These inocula either come from laboratory collections, such as INVAM (http://invam.wvu.edu/), or depend on commercial inocula.

**Commercial Inoculum**

The obligatory nature of the AM symbiosis for the fungal partner, because completion of their lifecycle depends on their ability to colonize a host plant, creates not only a severe constraint to its study but also hampers the possibility of large-scale inoculum production. So far, it is not possible to cultivate AMF axenically in large bioreactors as it has long been done for rhizobia and other symbiotic organisms. There are several difficulties associated with AMF inoculum production:

- the mandatory need of a suitable host plant to multiply the AMF;
- the time required for the plant to grow and the fungi to sporolate;
- the physical room required to cultivate the host plant;
- the several mechanical operations needed to collect and prepare the inoculum itself.

(*Continued*)

**BOX 3.1 (Continued)**

These constraints result in a very considerable cost of the final product and the impracticality of applying it on a large scale. To overcome the obligate biotrophic nature of AMF, the fungal propagules can be cultured in association with Ri T-DNA transformed hairy roots (Mosse and Hepper, 1975; Rodrigues and Rodrigues, 2015). Therefore one approach to reducing production costs is the *in vitro* mass production of inoculum associated with transformed roots (Adholeya et al., 2005), which requires carefully controlled growth conditions, with great emphasis on maintaining the effectiveness of inoculum and its genetic stability.

However, the problems of commercial AMF inoculum are not only linked to the price. Inocula are not very diverse, as they are mainly composed of AMF species that can easily sporolate, commonly *Rhizophagus irregularis* and *Glomus etunicatum*. Even though AMF lack absolute host–plant specificity there are host preferences, some degree of ecological specificity (McGonigle and Fitter, 1990) and AMF functional diversity (Munkvold et al., 2004), which is difficult to bring about with the limited diversity of commercial inoculum.

In the context of field application, the indigenous AMF populations can represent an additional challenge as they are often more abundant and better adapted to local conditions, leading to disappointing results of inoculation and limited persistence of introduced inoculum. Advances in ecology during the past decade have led to a much more detailed understanding of potentially adverse consequences of introducing species into a new habitat. However, there is little information available on the ecological consequences of inoculating with AMF, particularly given the limited knowledge of their basic biology and diversity (Abbott et al., 1995; Saito and Marumoto, 2002).

The quality of commercially available products, especially the guarantee that these are pathogen-free, the conditions required for storage before application, the most effective application methods and what is the appropriate inoculum for a particular application, are frequent concerns of potential users. Furthermore, information provided by suppliers about an inoculum can be deceiving, given that total counts of spores or propagules may be given, but only a fraction may be effective for a particular plant or under specific soil conditions.

An inoculant product is best used when there is reason to believe that the indigenous soil population of AMF is very small. This is common in nonagricultural applications, such as restoration of severely disturbed soils, degraded or unvegetated landscapes (e.g., highway embankments). It is also appropriated for plants that go through a nursery phase before being transplanted to the field with an optimized mycorrhizosphere. Most plants of interest in horticulture, tree-fruit culture, and forestry come into this last category (Azcón-Aguilar and Barea, 2015). Although inoculation with AMF has potential significance for sustainable crop production, including environmental conservation, current knowledge, and technologies limit the application for widespread use in many agricultural contexts.

typical field soil fertility levels (28 μg $NaHCO_2$ extractable P per gram of soil). The apparent contradiction between soil P availability and mycorrhizal effects may have more to do with the way colonization rate is assessed. Most of the time the increased root length, resulting when higher levels of P are available in soil, are not taken into consideration (see Section 2.1.3). These can result in the proportion of root length that contains arbuscules or hyphae appearing to decrease, even though the fungal partner continues to explore at the same rate. In the past AMF diversity has commonly been evaluated using classical trap cultures, rather than using the more exact molecular tools now available. Trap cultures only identify a relatively small number of the AMF present. Consequently, the earlier conclusion that AMF diversity is reduced by P addition might need to be revisited (see Section 2.2).

*AMF can enhance the N supply to plants by promoting acquisition of organic sources and may support mineralization. The ERM can become part of a common mycorrhizal network interconnecting pasture or intercropped species and permit direct transfer of N from legume to nonlegume plants. Nevertheless impacts of N may be less significant than those of P.*

Nitrogen (N) is also essential for plant growth. However, compared with the nutrient demand of the plant for growth, the contribution of AMF to plant P uptake is usually much larger than the contribution to plant N uptake (George et al., 1995). The mobility of $P_i$ is typically much less than inorganic N, partially explaining why the role of AMF in absorbing N may be of less significance than P. AMF are able to absorb both $NO_3^-$ and $NH_4^+$ and provide them to the host plant (Johansen and Jensen, 1996; Jin et al., 2005), and although they seem to absorb $NH_4^+$ in preference to $NO_3^-$ (Tanaka and Yano, 2005), the export of N from the hyphae to the roots and shoot may be greater following $NO_3^-$ uptake (Ngwene et al., 2013). Plant transporters specific to or induced by the symbiosis have been described for ammonium (Guether et al., 2009). AMF can also promote the acquisition of organic sources of N and to some extent promote mineralization (Hawkins et al., 2000; Hodge et al., 2001), something not possible for nonmycorrhizal plants. Hawkins et al. (2000) reported that mycorrhizal wheat absorbed between 0.2% and 6% of organic N in the form of N-glycine at low and high levels of N-application, respectively, although it was recognized that the amounts of N transported via hyphae contributed little to the N nutrition of the plant. Even so, the contribution from organic sources can be potentially valuable under N-limiting growth conditions for plants, associated with either scarcity or reduced mobility of N under water-limited conditions.

To determine the increase in N content of AM plants and quantify the contribution from the fungus, techniques based on the use of the $^{15}N$ isotope have been adopted (Azcón-Aguilar and Barea, 2015). Azcón et al. (2001) carried out one such experiment in a controlled environment to identify the proportion of plant N derived from fertilizer (%NdfF) in lettuce plants colonized by two different AMF species (*Glomus mosseae* and *G. fasciculatum*). N applications equivalent to 84, 168, and 252 mg $kg^{-1}$ of soil were used and P was also supplied to the plants. The authors observed that the application of increasing amounts of N did not change the percentage of root-length colonization by either AMF species tested. However, the total mycorrhizal root length (cm) was greater after inoculation with *G. fasciculatum* and was negatively affected by increasing amounts of N. With both AMF species the greatest value of %NdfF and fertilizer use efficiency was attained with the smallest N fertilizer application. *G. fasciculatum* reached the greatest value of labeled fertilizer utilization at 168 mg N $kg^{-1}$ soil, whereas with *G. mosseae* it was reached at 84 mg N $kg^{-1}$ soil. In this case the two species showed quite distinct behaviors. The results also indicate that mycorrhizal plants can regulate their uptake of N with respect to the amount of N in the soil. Although Miransari (2011a) recognized that large amounts of N fertilizer can significantly decrease N uptake by mycorrhizal plants, thereby affecting the nutrient management strategy, this is not independent of the circumstances and need for balance with other nutrients, such as P. Experiments focusing on separate as well as the combined effects of P and N indicate that root colonization was reduced, only when both elements were available in sufficient concentrations for the plants (Corkidi et al., 2002). This was confirmed by Blanke et al. (2005), who reported that N deficiency stimulated root colonization by AMF in a P-contaminated field site (up to 120 g $kg^{-1}$ soil). Therefore the N:P ratio might be an important factor determining AM development (Liu et al., 2000; Miller et al., 2002) (see Section 2.1.3).

The tripartite symbiosis between a legume host plant, AMF, and the N-fixing bacteria, collectively known as rhizobia, can affect the acquisition of N by the host plant. A synergistic effect between both microsymbionts and the host plant leads to the idea of a tripartite symbiosis (El-Hassanin and Lynd, 1985; Niemi and Eklund, 1988) and has been explained on the basis of a large P demand in the $N_2$ fixation process, which can be met through the symbiosis with the AMF (Stribley, 1987; Smith and Read, 2008) (see Section 5.1.1 and Box 5.1). However, there is now evidence that P uptake is not the main driver for the development of the tripartite symbiosis. In fact both symbioses share transduction pathways (see Box 5.1) and effects of the tripartite symbiosis (e.g., increased nodulation) can be observed within 10 days after plant emergence, when the seedling is still relying on P contained in the cotyledons (Goss and de Varennes, 2002; Antunes et al., 2006a).

Common mycorrhizal networks can interconnect component intercropped species or cultivars by extending AM mycelia from one plant root to another (He et al., 2003) and the direct transfer of N from legume to nonlegume plants mediated through AMF has been shown (Haystead et al., 1988; van

Kessel et al., 1985; Frey and Schüepp, 1992; He et al., 2003). The common mycorrhizal networks can also promote the transfer of nutrient from a decaying plant to an active one (Smith and Read, 2008). These interactions need to be taken into consideration as they may provide benefits for both agricultural and environmental management (see Section 6.3.2).

Just as in the case of P, the N uptake capacity, either for organic or inorganic forms of N, differs between AMF isolates, even those of the same species, depending on their background and ecotypic differences. Isolates from nutrient-poor sites commonly promote a larger uptake of N compared with that from a conventional agricultural field (Hawkins et al., 2000). After 18 years of N fertilizer application (up to 27.2 g $Nm^{-2}$ $year^{-1}$) in a grassland ecosystem and based on fatty acid profiles, Bradley et al. (2006) reported a decreased AMF diversity. On the other hand, using small subunit ribosomal DNA (SSU rDNA) analysis (see Section 7.1.1), Porras-Alfaro et al. (2007) observed that N-amended plots showed a reduction in the abundance of the dominant operational taxonomic units (OTU) but an increase in AMF diversity. This finding clearly indicates that shifts in AMF species composition and diversity can occur after N additions and may even result in more favorable conditions for AM symbiosis. For example, in a more diverse system the possibilities of successful combinations are greater and there may be a greater potential to take advantage of the opportunities associated with AMF functional diversity (Azcón-Aguilar et al., 1980). Clearly, specific environmental situations have to be carefully considered.

In addition to P and N, AMF are involved in the uptake of other nutrients. According to Clark and Zeto (2000) the information on the acquisition of macronutrient cations, potassium (K), calcium (Ca), magnesium (Mg), and sodium (Na) by AM plants have been relatively inconsistent in that increases, no effect, and decreases have all been reported. This range of effects is closely connected to different AMF species or isolates, the host plant and site characteristics, such as soil pH and the availability of other nutrients, especially P. Acidic soil generally contains only small amounts of these cation bases, which often become limiting to plant growth, but enhanced acquisition of K, Ca, and Mg has commonly been reported if plants are mycorrhizal. Even so, under these circumstances, liming can be an important strategy to adopt for increasing the effectiveness of AMF in the acquisition of these nutrients (Nurlaeny et al., 1996). Acquisition of sulfur (S) can also improve if plants become mycorrhizal but again AMF isolate, plant host, and site condition seem to be important determinants of the range of benefits obtained (Clark and Zeto, 2000).

Zinc (Zn) supply to plants depends on diffusion, as its mobility in the soil solution is very limited. Consequently the increased soil volume accessible to AM plants, as in the case of P, enhances Zn uptake positively and affects its concentration in various tissues of crop plant under particular environmental conditions (Ryan and Angus, 2003; Lehmann et al., 2014).

However, the impact may be less than reported for P because of the greater plant requirement for the latter. Gao et al. (2007) described a very strong interaction between Zn uptake by AM rice and plant genotype. They noticed that although AMF inoculated plants produced more biomass and took up more Zn than non-AM controls, AMF inoculation only significantly increased Zn uptake in those genotypes that had a small Zn uptake in the non-AM condition. The authors concluded that genotypes that are less efficient at taking up Zn in the absence of AMF show greater increases in efficiency when inoculated. Hajiboland et al. (2009) attributed this difference in genotypic behavior to a varying contribution of mechanisms for increased nutrient uptake in mycorrhizal plants, which depended on nutrient, nutritional status, and nutrient uptake efficiency of genotypes. These are all important traits that need to be kept in mind and therefore it is important to avoid making sweeping generalizations.

Large concentrations of boron (B) in soil may have negative effects on root AM colonization (Ortas and Akpinar, 2006) and B acquisition in AM plants seems to be relatively inconsistent, and it is also reported that AMF vary extensively in their ability to acquire or restrict availability of B (Clark et al., 1999).

*AMF can provide significant benefits to host plants by improving the supply of mineral nutrients that depend on diffusion for transfer to the root surface. At the same time, they can protect plants from toxic levels of metals.*

Reduction of manganese (Mn) absorption in AM colonized plants has been reported (Nogueira et al., 2004, 2007), these aspect may by particularly relevant in acid soils or other soils that have undergone temporary waterlogging, resulting in reduced soil oxygen sufficiently to convert sparingly soluble Mn oxides to the more soluble $Mn^{2+}$ form, easily leading to toxic levels of the minor nutrient in the soil. Nogueira et al. (2004) suggested a possible interaction with enhanced P absorption, which could increase plant tolerance to the internal concentration of Mn. Depending on the AMF species and the prevailing source of AMF propagules, with intact ERM granting better protection, the degree of resilience conferred by AMF against Mn toxic levels in the soil can vary (Nogueira and Cardoso, 2002; Brito et al., 2014; Alho et al., 2015). Also iron (Fe) acquisition by AM plants can be greatly affected by soil pH and the specific fungal symbiont. Amounts in the plant have been reported to be either enhanced or reduced, depending on the experimental conditions (Al-Karaki and Clark, 1998; Nogueira and Cardoso, 2002).

One potential consequence of increased uptake of nutrients is that the loss of nutrients to the environment could be reduced. There have been reports that the presence of AMF can result in large reductions of the $NO_3^-$

(Asghari and Cavagnaro, 2011), $P_i$ and $NH_4^+$ (van der Heijden, 2010) being leached, albeit the studies were carried out in very sandy substrates. Although there is evidence that both $NO_3^-$ and $NH_4^+$ are taken up by AMF hyphae, the reduction in $NO_3^-$ leaching could be more associated with greater $NH_4^+$ uptake reducing the amount undergoing nitrification or reductions in drainage from greater water uptake by AM plants. Bender et al. (2015) also investigated the impact of mycorrhizal plants on the loss of nutrients and less leaching of $P_i$, from two soil types having plants with AM than without. In neither an acid sand from a heathland nor a fine-textured soil under pasture did AMF have a significant effect on the leaching of $NO_3^-$. However, leaching of $NH_4^+$ was greater in the absence of AMF, although total losses were small in both soils. Mycorrhiza enhanced plant growth in the pasture soil but not in the acid sand, so as soil mineral N was not significantly affected by the presence of AMF, the mechanism responsible for the differences in leaching between mycorrhizal and nonmycorrhizal systems was unclear. Importantly, nitrous oxide fluxes from both soil types 24 hours after a pulse of $NO_3^-$ fertilizer were almost 50% less in the presence of AMF than in their absence. These apparent benefits from mycorrhiza could further contribute to reducing the impacts of agroecosystems on important environmental concerns; eutrophication of water resources and the increase of greenhouse gases in the atmosphere.

### 3.1.2 Defense Against Abiotic Stresses

Soil contamination resulting from both natural conditions and anthropogenic activities is a serious problem in many areas. Mainly by improving plant nutrient absorption and by influencing the fate of the metal and metalloid ions in the plant and soil, AMF can protect host plants against metal toxicity. The ability of AM plants to grow in heavily contaminated sites has long been reported; they can even be used in phytoremediation (Heggo et al., 1990; Ahmed, 2006; Yang et al., 2015). The symbiotic effectiveness of AMF varies with fungus and host plant species, soil properties and even AMF ecotypes (Turnau et al., 2001; Audet and Charest, 2006; Göhre and Paszkowski, 2006; Yang et al., 2015). It appears that a prolonged exposure to metals and metalloids can result in the development of some degree of tolerance in the AMF (Oliveira et al., 2001).

Saline soil contains sufficient neutral soluble salts to adversely affect the growth of most crop plants and presents a widespread problem, especially in regions with arid or semiarid climates. The salinity can also have detrimental effects on spore germination and the growth of hyphae, plant AM colonization, and formation of an effective symbiosis. There can also be a pronounced effect on AMF community structure across salt content classes (Juniper and Abbott, 2006; Krishnamoorthy et al., 2014). However, there are some salt tolerant plants (halophytes) that are highly colonized by AMF in their natural habitats (Bothe, 2012). Curiously, AMF spores in several saline soils consist of up to 80% of one single species, *Glomus geosporum*. In contrast, Bothe (2012) found that roots of halophytes were mostly colonized by fungi of the *Glomus intraradices* group, many of which had never been cultured.

Mycorrhiza inoculation can stimulate plant height, leaf area, root density, fresh, and dry plant weight of alfalfa under saline conditions (up to 150 mM NaCl) (Campanelli et al., 2013). In wheat AMF colonization can also result in increased uptake of N, P, K, Ca, and Mg, the presence of soluble sugars and free amino acids, accumulation of proline, as well as enhanced peroxidase and catalase activity (Talaat and Shawky, 2011). In both alfalfa and wheat, AM plants were found to have smaller sodium and chloride concentrations in their tissues compared to non-AM plants. Al-Garni (2006) claimed that in AM plants of *Phragmites australis* under salinity stress, amelioration of salt stress concentrations by the mycorrhizal association could be related to improved osmotic adjustment but independent of salt uptake of plants. Talaat and Shawky (2011) took a broader view and argued that mycorrhizal colonization can play a vital role in the mitigation of the adverse effects of salinity by improving the osmotic adjustment response in wheat, enhancing its defense system, and alleviating oxidative damage to cells. These authors stressed that AM are able to alter plant physiology in a way that empowers the plant to grow more efficiently on salt-affected land. Changes in plant water relations resulting from the formation of AM are discussed further in Section 3.1.4.

### 3.1.3 Defense Against Biotic Stresses

*AMF offer considerable potential in protecting host plants from biotic stress. The range of possible diseases that hosts can be protected against is considerable. Equally, the mechanisms effecting the protection are multifaceted, involving restricting invasion of the host and enhancement of defense responses.*

The persistent, careless use of chemicals to control plant pest and diseases has been discouraged owing to their toxic effects on nontarget organisms, including AMF (Abdalla and Abdel-Fattah, 2000), and other undesirable consequences in the environment. Many chemical substances, such as methyl bromide, have been phased out and the possibilities for fighting plant disease and predation through the use of synthetic pesticides are getting fewer.

AMF have been studied in a number of plant pathogen–host species combinations (Whipps, 2004) and, in general, are considered good allies in meeting the challenges to plant protection. They can provide host plants with some level of bioprotection against many agronomically relevant soilborne pathogens such as *Fusarium, Phytophthora, Pythium, Rhizoctonia, Sclerotinium, Verticilium,* and also various nematodes like *Heterodera, Meloidogyne, Pratylencus,* or *Radopholus* (Harrier and Watson, 2004). AMF can also be effective against root hemi-parasites, such as *Striga hermonthica,* as reported for tolerant sorghum plants by Lendzemo and Kuyper (2001). The protection conferred by AMF is specific to microbial isolate and host plant species (Pozo and Azcón-Aguilar, 2007; Sikes, 2010). This specificity seems to depend on a complex interaction between host plant, AMF, and pathogenic agent, not directly related to antagonism, antibiosis, or mycoparasitism processes (Harrier and Watson, 2004). Several mechanisms mediating the AMF–host plant interactions, operating together or separately, can cumulatively grant plant bioprotection (see reviews by Harrier and Watson, 2004; Wehner et al., 2010). These mechanisms include improved nutrient status of the host plant, anatomical, and architectural changes to the root system, competition for colonization and infection sites, changes in the total amount of carbon compounds released into the soil from a root system (rhizodeposition), changes to the microbial community in the rhizosphere and the activation of plant defense mechanisms, both local and systemically. The changes involved can collectively be communicated from one plant to another by the transmission of signals through the common mycorrhizal networks. Such networks appear to function as a plant-to-plant underground messaging service, whereby disease resistance and induced defense signals can be transferred between the healthy and pathogen-infected neighboring plants. This suggests that plants can "eavesdrop" on defense signals from the pathogen-challenged neighbors through this network and activate defense mechanisms before they themselves are attacked (Song et al., 2010). The priming of jasmonic acid-dependent defenses of AMF colonized plant roots can induce resistance to plant pathogens (Cameron et al., 2013), approximating to the supply of a vaccine for enhancing the immunological defense of the host plant. The role of microbial symbionts in immunological priming also raises the more general question of how hosts permit infection by beneficial symbionts while simultaneously discriminating against damaging pathogens (Clay, 2014).

The expression of all these mechanisms mediated by AMF–host plant interactions is certainly more effective in a well-established AM symbiosis and therefore the extent of AM plant colonization, when facing the stressor agent, is directly related to the level of bioprotection achieved (Diedhiou et al., 2003; Khaosaad et al., 2007; Sikora et al., 2008). The time required to develop a well-established AM colonization is dependent on the type of AMF propagules available. Lendzemo and Kuyper (2001) draw attention to the fact that soil disturbance can affect the integrity of the ERM network (one, and possibly the key propagule), thereby decreasing AMF colonization of the plant to be protected, and therefore providing a less effective protection. Furthermore, to establish an adequate level of AMF colonization, the prior-inoculation of plants at nursery level can enhance bioprotection of species, such as vine (*Vitis rupestris*) (Petit and Gubler, 2006) and olive (*Olea europaea*) (Castillo et al., 2006a; Kapulnik et al., 2010). Despite the range of difficulties associated with commercial AMF inoculum (see Box 3.1), this is where it can have important practical application. For crops seeded directly in the field that is certainly not the case, as an adequate level of colonization is needed before the crop faces the stressing agent and the type of propagules associated with such inocula are not the most effective to establish an early AMF colonization of the host plants.

Most of the studies on the bioprotective effects of AMF are based on inoculation of a single or very limited number of species, which is certainly not representative of the whole range of interactions between AMF and pathogens. Wehner et al. (2010) reasonably assumed that there could be functional complementarily among AMF species in regard to processes leading to pathogen tolerance or resistance, both among and within groups of implicated mechanisms. Therefore it could be expected that pathogen protection would increase with AMF diversity if a single species within an AMF assemblage provides a different benefit. Certainly, Jun-Li et al. (2010) achieved a better result in suppressing *Fusarium* wilt of cucumber using a consortium of AMF than a single species inoculum and Maherali and Klironomos (2007) reported reduced pathogen abundance using a taxonomically diverse set of AM fungal species in the same experimental setting.

### 3.1.4 Water Relations in Arbuscular Mycorrhizal Plants

*AMF colonization is more efficient in conditions with moist soil, which can mask the impact of an otherwise less favorable cropping regime. AM plants can better withstand drought.*

Although mycorrhizal effects on plant water relations are not as remarkable or consistent as those for P acquisition on host growth (Ryan and Ash, 1996), it is known that AM colonization affects water relations of plants, particularly during drought periods (Augé, 2001; Zhu et al., 2012). As with other aspects of the physiology of AM plants, it is important to distinguish direct effects of AM colonization on drought resistance from indirect effects resulting from changes in plant size and nutritional status induced by mycorrhiza formation. Various mechanisms have been identified as having responsibility for the drought alleviation of AM plants (Mickan, 2014). These include more effective scavenging of soil water (Sieverding, 1991; Allen, 2007), hormonal involvement (Krikun, 1991; Pedranzani et al., 2016) – especially in conjunction with the use of plant growth promoting rhizobacteria (Calvo-Polanco et al., 2016) or mycorrhiza helper bacteria (see Section 5.1.2);

protection from reactive oxygen species (ROS) generated under stress conditions (Ruiz-Sánchez et al., 2010); improved P nutrition (Fitter, 1988; Al-Karaki and Clark, 1998); enhanced photosynthesis (Zhu et al., 2012), improvement in soil structure caused by ERM and the associated improvement in soil moisture retention properties (Smith and Read, 2008; Augé et al., 2001) together with improved soil-root contact; stimulation of gas exchange through increase sink strength and possible effects on osmotic adjustment (Augé, 2001; Porcel and Ruiz-Lozano, 2004).

According to Atkinson (2004), relative to nonmycorrhizal plants, AM plants are apparently better adapted to their environment in terms of the rational use of water, since they use less water as the soil water potential decreases. Consequently rather than having access to more water, AM plants use the same quantity of water but over a longer period than non-AM plants. Liu et al. (2015) have given a practical demonstration of improved water use efficiency in AM poplar plants.

Hyphae from the ERM can contribute to water absorption. For example, in a split pot experiment with barley plants colonized by *Glomus intraradices*, Khalvati et al. (2005) reported that 4% of the water was removed from a compartment only accessible by the fungal hyphae. More recently, Ruth et al. (2011) claimed that, in a similar study, the AMF hyphae supplied at least 20% of total water uptake during a succession of drying cycles. However, some authors have argued that direct water transport through hyphae to host plants is improbable or of little significance in humid climates (Kothari et al., 1990; George et al., 1992). Augé et al. (2003) investigated the tolerance of bean (*Phaseolus vulgaris* L) leaves to dehydration. They found that direct and total effects of hyphal colonization of soil on both the leaf water potential, attained at the point of death from dehydration, and the corresponding soil water potential were greater than the effects due to colonization of roots, root density, soil aggregation, soil glomalin concentration, leaf P concentration, or leaf osmotic potential. Their results supported the assertion that external soil hyphae may play an important role in mycorrhizal influence on the water relations of host plants. These authors also reported that colonization of bean plants by a mix of AMF, collected from a semiarid grassland soil, resulted in a greater ability to withstand drought than did those colonized by *G. intraradices*. Quilambo et al. (2005), after testing a commercial inoculum, identified that it only had a positive effect on growth under well-watered conditions. Under drought conditions an indigenous Mozambican inoculum performed better as it was more adapted to the water-limited conditions of the region (Quilambo et al., 2005). Both these studies supported the view that a mixed AMF population, which was well adapted to the specific edaphic or climatic circumstances, would provide the most benefit.

Recent investigations by Kikuchi et al. (2016) suggest that the transport of polyphosphate granules, the form in which P moves through the ERM and the intraradical mycelium to the interfaces between fungus and host, occurs by mass flow. Furthermore these authors propose that the flow is driven by the transpiration of the host plant, acting as a sink. They show that introducing defective genes associated with the aquaporins that form pores for water movement across plasma-membranes, effectively slowed the movement of polyphosphates in the AMF *Rhizophagus clarus* HR1 and transpiration in the host plant *Lotus japonicus* L. These hypotheses contribute to an understanding of how water supply can be influenced by the AMF and the linkage between the water requirements of the host plant and the supply of P.

Experiments made by Karasawa et al. (2000a,b), with maize under various soil moisture conditions, showed that AM colonization improved with increasing soil moisture status, even when cultivated after a non-AM crop, thereby promoting AM formation, P uptake, and plant growth. The increase in AMF colonization with the increase in soil moisture status, despite the limited AM inoculum, suggests that greater soil moisture improves the efficiency of AMF colonization, promotes AM formation, and masks the influence of a less favorable AMF population in the soil. Consequently adequate moisture can reduce the influence of a negative cropping history.

## 3.2 CONSTRAINTS TO INTENTIONAL USE OF AMF IN AGRICULTURE

The range of direct and indirect benefits to the host plant by the association with AMF is vast, and it seems obvious to question: why are the outcomes not more often deliberately made available in cropping systems? In part, the answer is that the benefits granted by AMF are not always obvious or readily perceived by practitioners, making the intentional use of mycorrhiza in agriculture poorly developed and with very limited implementation. Abbott and Robson (1991) claimed that, for field crops, adequate management of AMF within the cropping system was missing and, despite the progress in research on mycorrhiza over the intervening years, the understanding of the functional ecology of AMF in agroecosystems is still poor.

> *The benefits granted by AMF are not always obvious or readily perceived by practitioners, making the intentional use of mycorrhiza in agriculture poorly developed and with very limited implementation. Our improved understanding provides a new impetus.*

Given the complexity of the AM symbiosis and all the interactions it can experience under natural conditions, it is unlikely that there will be a unique relationship between AM colonization and mycorrhizal benefit (Abbott et al., 1995). In reality, a low level of colonization does not necessarily mean that an AM symbiosis is worthless. Percentage colonization is almost the only calculation that has been made to quantify and compare the extent of host plant invasion, and measured as the proportion (%) of root colonized. However, it can be misleading about the extent of the symbiosis in a given system (soil, AMF, and host plant) and time, as it does not take into account the relative growth rate of both organisms under the prevailing conditions (Carvalho et al., 2015; see Box 2.2). In general, the effect of AM colonization on shoot behavior has often not been closely linked to the extent of the AM colonization of roots (Fitter and Merryweather, 1992; Miller et al., 1995a; Smith and Read, 2008), resulting in there being little relationship between crop yield and level of root colonization (McGonigle, 1988). In fact, good colonizers are sometimes inferior mutualists when compared to less infective species or isolates of AMF (Hetrick et al., 1993). Thus, when considering cultural traits, assessing the species composition of AMF communities may sometimes be more important than assessing levels of colonization or densities of spores in the soil (Johnson and Pfleger, 1992).

A strong common theme when going through the description of AM benefits (see previous section), was the different performance of distinct AMF species, or even isolates, and the importance of mixed inocula, highlighting the importance of a diverse AMF population for a greater receipt of AM benefits. More positive responses of plants to multiple AM fungal species may result from the great functional diversity and complementarily within AMF (Koide, 2000; Hart and Reader, 2002b; Munkvold et al., 2004; Maherali and Klironomos, 2007). Although not fully understood or controlled under cropping systems, it is unquestionable that the importance of a diverse AMF population has been well recognized (Bothe, 2012; Hawkins et al., 2000; Quilambo et al., 2005; Schweiger and Jakobsen, 1999). There is also plenty of evidence that indigenous AMF are usually effective in protecting plant hosts against biotic and abiotic stresses, even though these communities have been rarely considered in many of the studies reported (Whipps, 2004; Wehner et al., 2010). This is particularly the case when the effects of AMF on plants are more positive and the symbiosis occurs in a more realistic biotic context. Key facets include the potential colonization by multiple species of AMF associated with a diverse soil community, since beneficial fungi are more likely to be present in a mixed inoculum (Vogelsang et al., 2006). In a metaanalysis of context dependency in plant response to inoculation with mycorrhizal fungi, Hoeksema et al. (2010) suggested that plant functional group and inoculum complexity are relatively more important variables than P fertilizer application. The protection or improvement of soil AMF diversity should therefore be a matter of concern in the design of cropping systems, since it expands the possibilities of an effective AM symbiosis appropriate to the prevailing conditions. Currently for economic reasons and especially for high-input systems, the growth of monocultures or alternating between only two crops is frequent. Knowing that AMF and specific host plants have preferential associations (Torrecillas et al., 2012), the simplification of crop rotations leads to important changes in AMF diversity, not only in terms of the total population but also for the dominant species or isolates within that population, which unsurprisingly have negative consequences for AM benefits to the host plant.

As mentioned previously in describing the advantages to host plants made available by AM symbiosis, the temporal pattern of colonization is especially important, i.e., to have a well-established symbiosis from the beginning of the crop vegetation cycle. Even if diluted over time, this feature can still have repercussions for crop yield (Miller et al., 1995a). The type of AMF propagules available shapes how fast AM colonization takes place (see Section 6.2), and within the possible AMF propagule forms: spores, colonized root fragments, and ERM, it is the ERM that is the propagule which promotes an earlier and faster AM colonization (Martins and Read, 1997; Goss and de Varennes, 2002; Brito et al., 2014), even when relatively large levels of P (45 kg $ha^{-1}$) are available in the soil (Brito et al., 2013b) (see Section 5.1) or under field conditions (Brito et al., 2012a). Inversion tillage represents a severe threat to the AM symbiosis, which is aggravated in intensive cropping systems. The threat is not only because by disrupting ERM it interferes with a particularly important source of AMF propagule but also because it changes the AMF community structure (abundance and diversity) with consequences on AMF biodiversity. Ploughing (to more than 15 cm) hinders subsequent mycorrhiza formation by reducing propagule density in the rooting zone (Jasper et al., 1989a; Evans and Miller, 1990; Kabir et al., 1998b) and additionally it selectively interferes with different AMF, depending on their life and colonizing strategies, promoting, or impairing specific groups (Abbott et al., 1992; Brundrett et al., 1999; Klironomos and Hart, 2002). As a consequence reduction in the diversity of AMF communities in roots from distinct agroecosystems, specifically imposed by conventional tillage, has been reported in several studies (Alguacil et al., 2008; Schnoor et al., 2011). In a field experiment that had been under no-till for several years and assessing AMF diversity at soil level through DNA sequencing, Brito et al. (2012b) found that conventional tillage decreased AMF diversity by 40%.

*There is an urgent need for a more rational use of fertilizers to provide new opportunities for AM symbiosis to be integrated into nutrient management strategies.*

The excessive use of fertilizers is also frequently identified as a negative factor in the intentional use of AMF in cropping systems and AMF diversity, even if that is not always the case since some studies (Hoeksema et al., 2010) have shown the potential for vegetable producers to reduce their P fertilizer inputs and obtain comparable or superior yields with the help of AMF. However, the global P reserves suitable for the production of fertilizers are limited and might experience depletion within the next century (Cordell et al., 2009; Van Vuuren et al., 2010). In addition to the increasing cost of inorganic P fertilizer, which is manufactured from nonrenewable rock phosphate, ever-stricter agro-environmental guidelines, especially those associated with the quality of surface water resources, are encouraging vegetable growers to reevaluate and possibly reduce their use of P fertilizer (Elbon and Whalen, 2015). So too, the production of mineral N fertilizers is highly dependent on declining fossil energy resources (Vance, 2001). This reality requires an urgent need for a more rational use of fertilizers and will definitely offer new perspectives for the AM symbiosis to be integrated into nutrient management strategies for field vegetable crops.

It is commonly held that plant breeding programs, established under conditions of a high level of soil fertility, and ignoring AM symbiosis as a selection trait, have developed less mutualistic varieties. This is said to be one reason why formation of mycorrhiza by many crop plants is negligible. Although Turrini et al. (2016) found some confirmation of this idea in sunflower, it was not the case in wheat, since Hildermann et al. (2010) found no change in the ability of modern wheat cultivars to be colonized by AMF and Kirk et al. (2011) actually found that older cultivars produced significantly less AMF colonization and grain yield than modern cultivars.

*More positive responses of plants to multiple AM fungal species may result from the great functional diversity and complementarily within AMF. The protection or improvement of soil AMF diversity should be a key aspect of well-designed cropping systems.*

It is very important that a better common awareness and technical understanding of the AM symbiosis be developed. Many farmers are unaware of AM or that their benefits are within reach and can be obtained by the adoption of simple crop management techniques. The idea frequently held that the benefits granted by AMF have to be associated with the use of commercial inoculum can also be a constraint to their exploitation. A multiplicity of problems associated with commercial products, particularly the price and the poor biodiversity involved (see Box 3.1), often lead to deceptive results. Another negative aspect is that frequently technical and commercial staffs are not properly trained about the true facts surrounding the use of AM and consequently incorrect ideas about AM inoculum are promulgated, presenting it as a "miraculous" product that will allow dispensing with fertilizers while maintaining crop yields or even increasing them. On top of all this, guidance for inoculum application is frequently incorrect and it has also been known for inoculum to be sold as a suspension to be spread on plant leaves!

*The compatibility with local indigenous AMF and the persistence of introduced inoculum may be limited.*

## 3.3 CONCLUSIONS

Despite the considerable evidence of the wide range of benefits in crop production granted by the AM symbiosis, such as facilitated nutrient acquisition and protection of host plant from biotic and abiotic stresses, its intentional use in agriculture is far from being a reality. A number of factors contribute to this situation. These include the lack of knowledge of AMF biological and functional diversity and the impracticability of large-scale use of commercial inoculum due to its price and modest efficacy. Further constraints are closely related to crop management practices, which are still current in some regions of the world, such as simplified crop rotations or monocultures, soil inversion tillage, and indiscriminate use of fertilizers that together assist the decline of most favorable inoculum sources like intact ERM and also AMF diversity. Although the appropriate management of AMF within the cropping system has to contend with a limited understanding of the functional ecology of AMF in agroecosystems, boosting of indigenous AMF populations bearing in mind the preferential AMF–host plant associations in the crop sequence design and also the development and preservation of ERM in agricultural fields for a prompt AMF colonization of the crop, can be part of a more adequate management of AMF within the cropping system and fit in typical practices of conservation and organic agriculture (see Sections 2.1, 2.2 and Chapter 8).

Chapter 4

# Diversity in Arbuscular Mycorrhizal Fungi*

## Chapter Outline



Soil quality and ecosystem multifunctionality are determined by the diversity of species and functional groups present in the soil (Wagg et al., 2014) as well as the ability of the soil to respond to disturbances. Soil responses are influenced by the resistance and resilience of the soil microbial community (Griffiths and Philippot, 2013). The microbial population in soil is an essential component of the earth's biota, being important in essential ecological processes, such as soil structure formation, decomposition of organic matter and xenobiotics, and recycling of essential elements and nutrients (Madsen, 2005). Soil microbes also interact directly and indirectly with each other (Bardgett and Wardle, 2010) (see Section 5.1).

Microbial communities associated with plants have also been seen as an extension of the plant genome, being crucial to plant health and development (Mendes et al., 2011; Berendsen et al., 2012; Panke-Buisse et al., 2015). For example, the microbe–plant associations may improve stress tolerance,

* With Clarisse Brígido.


**Functional Diversity of Mycorrhiza and Sustainable Agriculture.**
**DOI: http://dx.doi.org/10.1016/B978-0-12-804244-1.00004-6**

disease resistance, and promote the biodiversity of plants in the ecosystem (Morrissey et al., 2004) through the deposition of nutrients, antibiotics, and plant hormones around the roots (de Vrieze, 2015). However, these microbial communities are in constant change, depending on several factors, including land use, the developmental stages of the different plant genotypes present, the nutritional status of the plants, and the presence of other organisms that may be beneficial, pathogenic or predatory on them or the plants (Kreuzer et al., 2006; Pineda et al., 2010; Chaparro et al., 2013; Nakagawa et al., 2014; Tsiafouli et al., 2015).

The symbiotic relationship between arbuscular mycorrhizal fungi (AMF) and their plant hosts is ancient and may have helped plants adapt from aquatic environments, in which nutrient resources are directly available, to terrestrial habitats, where nutrient-depleted zones rapidly develop after element absorption by roots (Sanders and Tinker, 1973; Corradi and Bonfante, 2012) (see Section 3.1). When AMF–plant symbioses are established, there are fungal structures growing inside plant roots as well as in the surrounding soil. The extraradical mycelium (ERM) is the part that grows outside the plant roots and acts as an extension of the root system, by increasing the absorption surface area, exploring greater soil volumes, and increasing the lifespan of absorbing roots (Smith and Read, 2008). The intraradical mycelium grows mainly between the cells of the root and in the cortex forms highly branched structures (arbuscules) inside the cell walls, where the exchange of nutrients from fungus for carbohydrates from the plant occurs (Box 5.1). Not only are the fungi able to provide the host with mineral nutrients but they also provide additional benefits in the form of protection against a number of environmental stresses, biotic and abiotic (Smith and Read, 2008; Clark and Zeto, 2000; Turnau and Haselwandter, 2002) and by inducing changes in the root system architecture and in root exudates (Norman et al., 1996; Pivato et al., 2008). These changes can have direct impact on the microbial community in the rhizosphere, including beneficial microbes and root pathogens.

The variation in the soil microbial population resulting from impacts of changes in the environment, the interaction with plants and other microorganisms all contribute to the biodiversity in soil and hence can influence plant productivity. For AMF and host plants these same factors promote diversity but there is also diversity induced by the interactions of the different plant–microbe combinations as well as those with other soil organisms.

*AMF, as an important part of the soil biota, contribute to ecosystem formation, stabilization, and thereby provide links between plants and soil-based ecosystem services. Important for this is the functional diversity within the Glomeromycota – the phylum to which they all belong – and that stems from differences between taxa, as well as those within species, and those resulting from the interactions between an AMF and its host plant.*

## 4.1 ECOLOGICAL ROLES OF ARBUSCULAR MYCORRHIZAL FUNGI

The symbiotic association between different AMF and plants have several ecological consequences that can be summed up as follows: (1) enhancing uptake of poorly mobile ions, (2) developing and improving soil structure and structural stability, (3) increasing plant community diversity, (4) improving rooting and plant establishment, (5) increasing soil nutrient cycling, and (6) enhancing plant tolerance to biotic and abiotic stress (Smith and Read, 2008). There is evidence that colonization rates, growth of ERM, and the efficiency of phosphorus (P) uptake differ according to fungal species, as does mycorrhizal-specific gene expression in the host plant (Johnson et al., 1997; van der Heijden et al., 1998a). However, experimental evidence, based on hyphal growth and spore production, also shows that considerable genetic and phenotypic diversity exists within the same population of the AMF *Glomus intraradices* (Koch et al., 2004).

There is considerable evidence for a marked diversity among AM fungal communities belowground. However, as indicated previously, the AMF diversity is influenced by several factors, such as the host-plant species diversity, soil, and season, separately or in combination (Smith and Smith, 2012). For example, arable soils, subject to large inputs for crop production, commonly have little AMF diversity and contain only a few species, in contrast to perennial communities, such as natural grasslands or tropical forests (see Sections 2.1 and 2.2), which contain complex AMF communities with much greater diversity (Clapp et al., 2001). It has been suggested that diversity of AMF in the soil influences significantly that of the plants and the productivity of an ecosystem (van der Heijden et al., 2008a; van der Heijden and Horton, 2009), being responsible for species-specific feedbacks that influence both above- and belowground community composition (Bever, 2003). Despite theoretical support for a direct association between soil microbial diversity and plant diversity (Hooper et al., 2000), results from empirical studies have been inconsistent (van der Heijden et al., 1998b; Landis et al., 2004; Vogelsang et al., 2006; Thoms et al., 2010; Peay et al., 2013; Hiiesalu et al., 2014; Antoninka et al., 2011; Lekberg et al., 2013; Öpik et al., 2008; Öpik et al., 2010). Several explanations may account for differences found between the various investigations involving AMF. In some studies, the number of fungal or plant taxa involved was relatively small. In other cases, determination of AMF diversity was based on spore identification, which can underestimate some AMF taxa that do not establish in cultures or rarely produce spores (Sanders, 2004). Furthermore, some spores present in soil may not be produced by the AMF taxa colonizing the plant roots (Varela-Cervero et al., 2015). A further confounding finding from a study, conducted by Hiiesalu et al. (2012) in native grassland, was that the number of plant species detected belowground increased with soil fertility, one of the main determinants of habitat productivity, whereas species richness aboveground declined. The authors suggested that the decline in aboveground annual species might have resulted from exclusion owing to asymmetric light competition, with tall plants having a greater advantage over shorter ones. In contrast, perennial species could have remained dormant in belowground structures until more amenable soil conditions became re-established. Persistence of underground diversity would be supported by patchiness of fertility, which could be exploited more equitably by roots of different plants. In a following study, Hiiesalu et al. (2014) confirmed their earlier results and showed that the number of AMF taxonomic units increased with the richness of plant diversity, with the correlation being greater with the plant richness belowground. However, species diversity of AMF was negatively associated with plant biomass, whether based on above or belowground biomass. Efforts to clarify the links between the AMF diversity and plant productivity have established that genetically distinguishable AMF isolates, even those from the same species, have different effects on their host plants (Munkvold et al., 2004; Koch et al., 2006).

The degree of plant protection has been shown to correlate with the diversity of the AMF population present in colonized roots (van Tuinen et al., 1998b; Tian et al., 2013a). Moreover, several studies have indicated that protection against both biotic (Thygesen et al., 2004; Lax et al., 2011; Campos et al., 2013) and abiotic stresses (Kothari et al., 1991; Oliveira et al., 2006; Seguel et al., 2013) are dependent on the AMF functional diversity. Overall, there are two critically important aspects that have to be resolved. Firstly, the relationship between taxonomic and functional diversity in AMF and secondly, how well differences in genetic variation can be used to predict changes in productivity resulting from all the benefits to the host plants.

## 4.2 BASIS OF FUNCTIONAL DIVERSITY IN ARBUSCULAR MYCORRHIZAL FUNGI

The identification and classification of AMF along with other plant and animal groups have undergone major changes since the turn of the century, not least as a result of the advances in molecular techniques. The same advances have also allowed the factors that contribute to variation in the diversity of AMF to be studied.

### 4.2.1 Taxonomy of Arbuscular Mycorrhizal Fungi

It is generally agreed that all AMF are affiliated taxonomically into one monophyletic group within the fungi, the phylum Glomeromycota (Schüßler et al., 2001b). Previously, a large number of the taxonomic names at different levels within this phylum were not universally accepted. To resolve this issue the classification of Glomeromycota was revised and a "consensus" classification established that provides a framework for additional original research focused on clarifying the evolutionary history of AMF (Redecker et al., 2013). At the present, four orders are recognized, the Glomerales, Diversisporales, Archaeosporales, and Paraglomerales (Schüßler et al., 2001b; Walker and Schüßler, 2004). Within these orders, eleven families and 22 genera are characterized, which incorporate more than 220 different AMF species (Schüßler and Walker, 2010), identified on the morphological characteristics of the spores (Krüger et al., 2012; Öpik et al., 2013). Most of the species described belong to the genera *Glomus*, *Acaulospora*, *Scutellospora*, and *Gigaspora*. However, a number of groups of uncertain phylogeny have not yet been formally reclassified, in part because morphological identification of AMF based on spores can be limited by their simple structure and relatively few morphological characteristics, restricting identification to the family level (Merryweather and Fitter 1998). Further complications in classification arise from the difficulty of maintaining spores collected from the field in their original condition and the fact that several species of AMF may not produce spores under laboratory conditions (Lee et al., 2013). Consequently, there are problems in defining taxonomic boundaries between and within species, based on spore characteristics, and hence the true diversity of AMF in an ecosystem cannot be determined. The development of molecular techniques has allowed the discrimination at species or isolate-level of active root-colonizing members of the Glomeromycota (Redecker, 2002; Redecker and Raab, 2006).

The recognized role of these plant symbionts in the ecosystems led to the search for information regarding their ecology and geography (Fitter, 2005; Chaudhary et al., 2008) and became a crucial research area in the last 20 years. The application of molecular tools (see Chapter 7) for identification of Glomeromycota from environmental samples has increased considerably in the last decade and provided data for several ecosystems and regions around the world (Öpik et al., 2006; Helgason et al., 2007; Öpik et al., 2009; Varela-Cervero et al., 2015). Such studies have indicated that a large number of groups, based on the presence of common base sequences (phylogroups), are present in natural ecosystems and the AMF diversity in plant roots collected from the fields is unexpectedly large (Helgason et al., 2002; Becklin et al., 2012; Varela-Cervejo et al., 2015). However, for most of the phylogroups detected, no similar sequences were obtained from cultured reference species, demonstrating that their relationships with known species are unclear (van der Heijden et al., 2008a; Rosendahl, 2008). All these results are consistent with the view that overall AMF diversity has been greatly underestimated (Wang and Li, 2013). For example, the open-access database, called *MaarjAM* (Öpik et al., 2010), which stores publicly available Glomeromycota sequence data with the associated metadata and provides an overview of existing information about the ecology and geography of Glomeromycota, already accounts for 352 different phylogroups or virtual taxa (VT). It is expected that this number will increase as more environments are investigated.

AMF are able to colonize a large number of host-plant species successfully (Smith and Read, 2008). However, although they do not show strong host specificity, recent ecological studies on the diversity of AMF associated with different plants have confirmed the view that there is a strong preference shown by host plants for specific AM fungal genotypes (Croll et al., 2008; Scheublin et al., 2004; Öpik et al., 2009; Lekberg et al., 2013). Thus, molecular methods have not only provided important information on the natural diversity of AMF but may make it possible to overcome the limitations resulting from the need to characterize spores.

### 4.2.2 Diversity of Arbuscular Mycorrhizal Fungi Related to Growth Habit

Although all AMF can colonize roots and their rhizosphere soil simultaneously, even spreading out from roots over several centimeters in the form of ramified filaments (Nasim, 2010), species of AMF differ in their mode of colonization and capacity to form hyphae in soil and inside the root (Abbott et al. 1992). For example, AMF genotypes differ in hyphae production, viability and structure, even those collected from the same location (Sanders et al., 1977; Abbott and Robson, 1985; Giovannetti et al., 2001; Koch et al., 2004). Chagnon et al. (2013) used an ecological adaptive strategy approach (Grime, 1979) to consider how growth characteristics could inform the life strategy of different AMF families. Isolates of the Gigasporaceae were characterized by the formation of large hyphal densities in soil, late production of spores in the growing season, and providing greater nutritional benefits to hosts. Such characteristics were consistent with a strongly competitive life strategy. AMF able to cope with disturbance, Ruderals, typified by isolates of the Glomeraceae, were expected to show faster growth rates, greater intraradical colonization, develop spores after a short nonreproductive phase and only produce a small density of hyphae in soil. In contrast, stress-tolerating AMF would have slow growth rates and form long-lived mycelium that showed resistance to acid soil conditions and cold temperature. These characteristics were present in isolates of the Acaulosporaceae (see also Section 7.2). AMF species differ also in the amounts of ERM produced and consequently in their ability to take up and supply phosphorus (Helgason et al., 2002; Jakobsen et al., 1992a; Jansa et al., 2005; Munkvold et al., 2004). The variation in ability to take up and supply phosphorus to plants also extends to nitrogen and applies to isolates of the same species (see Section 7.2). It has been associated with the expression levels of genes coding for P-transporters and N-assimilation (Maldonado-Mendoza et al., 2001; Govindarajulu et al., 2005).

### 4.2.3 Interaction Between the Genotypes of Fungi and Host Plants and the Diversity of Arbuscular Mycorrhiza

The efficacy of the symbiotic interaction between AMF and plants is dependent on factors related to the genotypes of both fungal symbiont and host plant (Jakobsen et al., 1992a,b; van der Heijden et al., 1998a; Smith et al., 2000; Burleigh et al., 2002; Munkvold et al., 2004; Avio et al., 2006). Consequently, differences in essential functions, such as colonization rate, growth of the ERM, efficiency of phosphorus (P) uptake by AMF, and mycorrhizal-specific gene expression within different AMF species would be expected to have repercussions on the efficiency of the AM symbiosis and result in variation in growth responses of host plants. There is good evidence that genetically different AMF isolates vary in the impact they have on plant growth and this variation in functionality depends on the host plant (van der Heijden et al., 1998a; Maherali and Klironomos, 2007; Angelard et al., 2010). Although it might be expected that functional differences would be more evident among distinct phylogenetic groups, some authors observed functional variation even between isolates belonging to the same species group. For example, Hart and Klironomos (2002) showed that the variation in growth response of *Plantago lanceolata* was considerable in treatments involving different AMF genera but small in treatments containing different isolates from the genus *Glomus*. Similarly, Hart and Reader (2002) obtained evidence that substantially different benefits to the host plants only resulted from colonization by contrasting families of AMF. However, Munkvold et al. (2004) found that high functional diversity existed within a single AMF species with respect to P-uptake and plant growth stimulation. Moreover, Avio et al. (2006) detected both interspecific and intraspecific functional diversity in *Funneliformis mosseae* and in *Rhizophagus intraradices*. For each of the species the two isolates investigated affected plant P content or the P and N concentration differentially, although isolates of *F. mosseae* demonstrated more variability than did isolates of *R. intraradices*. Angelard et al. (2010) also working with genetically different AMF isolates of *R. intraradices* reported functional differences among the AMF isolates in their effects on host-plant growth.

All these studies supported the view that considerable functional diversity existed within AMF species and that the variation within one AMF species could be bigger than that between species or even genera. Nevertheless, phylogenetic characteristics of AMF functioning has also been shown to be somewhat conserved (Hart and Reader, 2002; Powell et al., 2009; Maherali and Klironomos, 2012; Chagnon et al., 2013). There is also strong evidence that the efficacy of an AM symbiosis can depend on the association between the plant species and AMF involved. For example, Klironomos (2003) reported that after testing more than 60 plant species, there was never a particular AMF isolate that was always associated with the largest impact on growth whatever the plant species. It was also notable that the greatest variation in growth effects was found when native plants were inoculated with AMF from the same location. Several studies on the interactions between AMF and their host plants have shown that the plant genotype not only had a significant effect relative to the compatibility with AMF (Heckman and Angle, 1987; Estaún et al., 1987; Boyetchko and Tewari, 1995) but also influenced the physiological interactions leading to differences in plant growth (Graham et al., 1982; Azcón et al., 1991; Monzon and Azcón, 1996; van der Heijden et al., 1998a). Moreover, differences in the host plants' responses to a specific AM fungus can be found at the level of plant species (Monzon and Azcón, 1996) or cultivars (Mao et al., 2014). Schoeneberger et al. (1989) reported a differential effect between *Trifolium subterraneum* and *Lotus pedunculatus* Cav. on growth and uptake of P, following inoculation with *Gigaspora margarita* Becker and Hall. An improvement in P uptake took place in *T. subterraneum*, but not in *L. pedunculatus*, which was attributed to differences in root architecture between the two host plants. Differential response to inoculation with the same assemblage of AMF can even be found between different genotypes of the same plant species (Hetrick et al., 1993).

The evidence for a large genetic diversity within the Glomeromycota coupled with inter- and intraspecific variability in function, with both these sources of variation being influenced by the interaction between the colonizing AMF and the host plant, is consistent with there being considerable functional diversity between arbuscular mycorrhiza.

## 4.3 FUNCTIONAL DIVERSITY ASSOCIATED WITH HOST-PLANT BENEFITS

AMF represent a considerable fraction of the total living soil biomass (see Box 1.3) and provide key links between plants and soil-based ecosystem services, such as nutrient cycling structural resilience and water supply (Klironomos et al., 2011). They thus form an important link between biodiversity and ecosystem functioning (Read, 1991). Although the functional benefits provided by AMF are well established, the variability in their efficacy is an important aspect when managing diversity. Our understanding is largely dependent on the introduction of inoculum (see Box 2.1) with two or more AMF isolates under controlled environments. Yet it has been suggested that AMF species complement each other by occupying different niches when they simultaneously colonize a root system (Koide, 2000; Maherali and Klironomos, 2007). In addition, since microbial diversity is influenced by many environmental factors (Gans et al., 2005; Wall et al., 2010; Cardinale et al., 2011; Bodelier, 2011; Bardgett and van der Putten, 2014; Tedersoo et al., 2014), it would be expected that the relationship between AMF richness and plant biomass production not only depends on the identity and functional attributes of the organisms involved, but also depends on the environmental conditions, especially the availability of limiting nutrients, under which they interact (Klironomos, 2003; van der Heijden et al., 2008a,b).

*Key benefits of the arbuscular mycorrhiza symbiosis can be summarized as the acquisition of plant nutrients, the protection from biotic and abiotic stresses, and the enhancement of soil physical conditions by formation and stabilization of soil structure. The diversity of AMF in relation to these functions is considered critical for the long-term stability of agroecosystems under climatic and anthropogenic change.*

### 4.3.1 Acquisition of Mineral Nutrients

Uptake of P, the length of the ERM, and the combined architecture of hyphae and roots have all been suggested as possible contributory mechanisms responsible for differential growth responses of plants colonized by AMF (Helgason et al., 2002; Jakobsen et al., 1992a; Munkvold et al., 2004). For example, inter- and intraspecific differences in AM fungal efficiency and the differential increases in P and N supply to host plants have been attributed to phenotypic and functional properties of the ERM (Ames et al., 1983; Jakobsen et al., 1992a,b; Mäder et al., 2000; Hodge et al., 2001; Smith et al., 2004). Moreover, differences between AMF in P uptake characteristics and ERM length might also contribute to the ability of a variety of naturally occurring plant species to coexist (van der Heijden et al., 2003; Oliveira et al., 2006). For example, plant species with a high level of mycorrhizal dependency exhibited greater benefits from AMF species with longer lengths of ERM, whereas AMF species with the smallest ERM length benefited plant species that were less dependent on mycorrhiza (Oliveira et al., 2006). The ERM allows the host plant to have access to a greater quantity of water and soil minerals required for its growth. The effective surface area for absorption is extended by the ERM beyond the zone of depletion around the roots that develop as nutrients are taken up. (George, 2000; Dodd, 2000).

The combination of impacts on nutrient acquisition and improvement in soil structure would suggest that plants colonized by Gigasporaceae isolates would have access to a larger soil volume and more nutrients but less benefit would result for hosts with roots extensively colonized by intraradical mycelium of isolates from the Glomeraceae but having sparser hyphal density in the soil (Maherali and Klironomos, 2007). However, a direct relationship between the amount of ERM and plant benefits is not always evident. For example, Thonar et al. (2011) assessed the diversity in P acquisition strategies among three species of AMF, *R. intraradices*, *Claroideoglomus claroideum*, and *Gi. margarita* and found that *R. intraradices* and *C. claroideum* were more efficient in providing P to roots, whereas *Gi. margarita* provided rather little P to the plants, despite the formation of a dense mycelium network close to the roots. Such results could be explained if the efficiency in uptake of nutrients, mainly P, by the ERM of the fungus in mycorrhizal plants, were dependent on the architecture of the external hyphae and hence on the soil volume exploited. Friese and Allen (1991) ascribed different functions, such as spread, infection or nutrient absorption to the hyphae of the ERM, all likely to be fungal specific, so competition between AMF together with differences in the spatial distribution of hyphae away from roots (Jakobsen et al., 1992a) could influence the efficiency of nutrient supply to the host. In addition, AMF dependency of plants, which is ascribed to host physiology, may account for differences in their response to specific fungal partners. In spite of the evidence for involvement of the exchange of phosphorus and photosynthate between AMF and plants, the significance of colonization rates and growth of ERM, and the knowledge that mycorrhizal-specific gene expression differs among fungal species, participation of other mechanisms in determining AM functional diversity is not clearly established (Helgason et al., 2002; Jakobsen et al., 1992a,b; Jansa et al., 2005; Munkvold et al., 2004). Diversity in ability to take up and supply phosphorus and nitrogen to host plants is also found among isolates of the same species and has been associated with the occurrence and activity of P-transporters and N-assimilation physiology (see Section 4.2.2). Mensah et al. (2015) studied the effect of 31 different isolates from 10 AM fungal morphospecies on P and N nutrition of *Medicago sativa* and found that performance of these isolates depended on the phosphate metabolism of the fungus. An important factor was that high-performing isolates belonged to different morphospecies and genera, which indicated that ability of AMF to contribute to nutrient acquisition, were widespread within the Glomeromycota. The AM function also depended on the root architecture of the host plant. Sikes et al. (2009) found that species of *Gigaspora* were more efficient than species of *Glomus* in promoting P acquisition and plant growth in *Allium cepa*, which has simple root system, whereas in *Setaria glauca* (L.) P. Beauv. with a complex root system, there were no differences in the P acquisition promoted by species of either genera. The complexity of the functional differences among AMF in promoting plant growth and nutrient acquisition also results from interaction with the gene expression of the host plant. Burleigh et al. (2002) found that the root expression of inorganic P ($P_i$) transporters (MtPT2 and Mt4) in the roots of *Medicago truncculata* depended on the AMF species. Colonization by *Glomus mosseae* resulted in the greatest reduction in expression of these genes and the highest level of P uptake and plant growth. In contrast, *Gigaspora rosea* colonization caused the highest level of the expression of these two genes together with the smallest P uptake and plant growth. However, with *Lycopersicum esculentum* as host plant, root expression of two $P_i$ transporter genes (LePT1 and TPSI1) and P acquisition was little affected by the same AMF species. Therefore the functional diversity of AM symbiosis in relation to P acquisition resulted from a complex combination of the architecture of the ERM, the expression of AMF genes coding for $P_i$ transporters together with the architecture of the root system of the host plant, and the interaction between the host plant and the fungus on the expression in the roots of $P_i$ transporter genes.

### 4.3.2 Protection Against Abiotic Stresses

Assemblages of AMF are recognized for helping plants to thrive in hostile environments (Barea et al., 2011). This is of fundamental importance because adverse conditions, particularly exacerbated by global climate change, generate a great array of stresses that affect the stability of both natural and agricultural ecosystems (Compant et al., 2010). Since abiotic stresses are common and limit crop productivity worldwide, the ability to exploit and manage AMF symbioses and thereby increase host tolerance to such stresses (Barea et al., 2011; Jeffries and Barea, 2012), could provide an alternative strategy for overcoming these environmental constraints to crop production.

There are numerous reports of the benefits to plants under abiotic stress provided by AMF, including tolerance to drought (Augé, 2001, 2004; Ruiz-Lozano, 2003; Miransari, 2010; Augé et al., 2015), salinity (Evelin et al., 2009; Miransari, 2010; Porcel et al., 2012), heavy metal and metalloid toxicity (Cornejo et al., 2013; Brito et al., 2014; Meier et al., 2015; Alho et al., 2015), heat (Compant et al., 2010), cold (Charest et al., 1993), industrial effluents (Oliveira et al., 2001), biocide treatment (Heggo et al., 1990), slurry application (Chistie and Kilpatrich, 1992), sulfur dioxide fumigation (Clapperton et al., 1990), wild fire recovery (Puppi and Tartaglini, 1991), or osmotic stress (Ruiz-Lozano, 2003). Here we only focus on those that are relevant to agricultural systems.

*In the case of manganese toxicity, differences in AM functional diversity in wheat and subterranean clover resulted from the interaction between indigenous AMF and two intermediate hosts.*

Different symbiotic performances from associations of various AMF and host plants have been observed under drought (Augé, 2001) and manganese toxicity (see Sections 3.1.1, 5.1.1). Mycorrhizal plants may differ in their response to stress due to the functional diversity of AMF, depending on the interaction between AMF, plants, and environmental conditions. In fact there are consequences for plants differing in their response to certain species of AMF, or if they vary in the level of dependence on the symbiosis. This has been demonstrated by unforeseen effects on plant communities that can occur from changes in the composition of the AMF present (van der Heijden et al., 1998a; Antoninka et al., 2009; Sun et al., 2013). Increasing AMF diversity in the soil community might increase the chances of establishing an AM with a fungal species that is more efficient for plant growth (Santos et al., 2008). Although it is recognized that there are plant–AMF species combinations more beneficial to host plants than others (Klironomos, 2003; Kiers et al., 2011; Walder et al., 2012), the specificity of plants in their association with particular AMF species might also limit the AMF association best fitted for a given environmental stress. For example, despite the large pool of AMF species present in soil, a greater growth of wheat and clover, under excessive levels of manganese, was only obtained when the composition of the AMF communities changed due to the presence of an intact ERM previously developed by *Ornithopus compressus* or *Lolium rigidum*. Plant growth was particularly enhanced when the *Developer plant* was *O. compressus* rather than *L. rigidum* (Brito et al., 2014; Alho et al., 2015), suggesting that the effect might be associated with functional consequences of the different AMF diversity imposed by these plants. Analysis of the communities of AMF present in the plant roots revealed preferential symbioses within an AMF population for a particular plant group. The AMF communities in roots of the plant species belonging to the Fabaceae (*O. compressus* L. and *T. subterraneum* L.) family differed from those in the Poaceae (*L. rigidum* G. and *Triticum aestivum* L.) family, when spores and colonized root fragments were the main propagule source. The *Rhizophagus* sp. OTU 01 was significantly more frequent in the two legumes than in the two nonlegumes (>58% and <34%, respectively). In contrast the uncultured Glomeraceae OTU 02 and uncultured *Claroideoglomus* sp. OTU 03 were preferentially found in the nonlegumes, on average accounting for more than 49% of the AMF community in these hosts but for less than 16% of the communities isolated from the roots of legumes. However, when intact ERM associated with the first plant type was present in soil and functioned as the preferential AMF propagule for colonizing the second plant in the sequence, such as when an intact ERM associated with Poaceae species is present in the soil before the plantation of Fabaceae species or vice versa, the AMF community composition present in the roots of the following crop becomes more similar to those in the first plants. For example, the presence of an intact ERM associated with *O. compressus* induced significant changes in the AMF community present in the wheat roots mainly due to the significant augmentation of colonization by *Rhizophagus* sp. OTU 01 from 4% to 70% of the AMF community and reductions in the colonization by the uncultured Glomeraceae OTU 02 and uncultured *Claroideoglomus* sp. OTU 03 from 81% to 4% of the AMF community. Alternatively, the presence of an intact ERM associated with *L. rigidum* greatly increased colonization by the uncultured Glomeraceae OTU 02 and uncultured *Claroideoglomus* sp. OTU 03 from 8% to 58% of the AMF community in the subterranean clover roots. In other words, when ERM was the preferential inoculum form, the AMF community present in the roots of the second plant matched that present in the first crop, whichever host families were involved (Brígido et al., 2017). These results suggest that within the indigenous AMF community there was functional diversity available for protection against Mn toxicity and that could be managed by the choice of the Developer plant and the preferential AMF inoculum source.

Under contrasting conditions of Al toxicity (with and without), Clark et al. (1999) showed that the relative benefits conferred by five different AMF species (*Glomus clarum*, *Glomus diaphanum*, *Glomus etunicatum*, *Glomus intraradices*, *Gigaspora albida*, *Gi. margarita*, *Gi. rosea* and *Acaulospora morrowiae*) changed, although some species were beneficial in both cases. Two important conclusions can be drawn from this study. The first is that AMF functional diversity in protection of plants against Al toxicity exists at least at species level. The second is that AMF functional diversity depends on the benefits the host plant requires. In the absence of Al toxicity the AMF species that enhanced nutrient acquisition were the ones that promoted better plant growth. However, when Al toxicity could impair plant growth the most beneficial AMF were those more able to exclude Al from the plant.

Protection against water stress by different AMF species seems to be related to specific physiological ($CO_2$ fixation, transpiration, water-use efficiency) and nutritional (P and K) mechanisms according to the fungus involved in the symbiotic association (Garg and Chandel, 2010). Marulanda et al. (2003) tested the effect of six different AMF *Glomus* species (*G. coronatum*, *G. intraradices*, *G. claroideum*, *G. mosseae*, *G. constrictum*, *G. geosporum*) on the growth and water extraction by lettuce (*Lactuca sativa* L.) under water stress. There was a considerable functional diversity in the effects of different AMF. The differences in relation to nonmychorrizal control ranged from 0.6% of volumetric soil moisture for plants colonized by *G. mosseae* and up to 0.95% in plants colonized by *G. intraradices*, with the differences being associated with the size of the ERM. Coinoculation of AMF strains could improve plant growth under water stress (Augé et al., 2003; Quilambo et al., 2005; Bompadre et al., 2013).

Salt-affected soils occur in more than 100 countries and their worldwide extent is estimated at about 1 billion ha (FAO, 2015) with the likelihood that they will increase in the future. Tian et al. (2004) studied the effect of two isolates of *G. mosseae* in the protection of cotton plants to excessive levels of salt in the soil. One isolate was collected from a nonsaline soil and the other from a saline soil. There were significant differences between the two isolates, with the isolate collected from nonsaline soil providing the best protection to the host plant. The results of Liu et al. (2016), reporting significant protection from salt stress of cotton and maize by native AMF, appear to be in direct opposition to those of Tian et al. (2004). Furthermore, the results presented by Bothe (2012), which showed that 80% of spores in a saline soil consisted of a single species, *G. geosporum*, whereas the roots of halophytes were mostly colonized by fungi of the *G. intraradices* group. This raises the question whether *G. intraradices* rather than *G. geosporum* was the major player that conferred salt tolerance to the halophytes. Bothe (2012) concluded, "there are indications that AMF may alleviate salt stress in plants, but the results published often do not appear to be convincing. An AMF isolate or a mixture of fungi that confers salt tolerance to plants might have enormous potential applications." Nevertheless, there are indications that functional diversity within AMF exists for the protection of plants against salt stress and could be used to improve the contribution of AMF.

Despite the general presence of AMF in harsh environments and the important potential of these symbiotic fungi to protect plants against several stresses, the mechanisms responsible for the increased plant tolerance to stress are yet to be fully elucidated. Recently, several authors have reviewed in depth the possible mechanisms responsible for the increase plant tolerance (Miransari, 2011b; Porcel et al., 2012; Seguel et al., 2013; Wu et al., 2013; Cicatelli et al., 2014).

Given the potential influence of AMF on plant responses to climate change, the direct effects on AMF and the consequences of abiotic stresses for the symbiosis cannot be ignored. For example, salinity can adversely affect the colonization capacity of the fungi, the germination of spores, and the growth of fungal hyphae (Juniper and Abbott, 2006; Hajiboland, 2013). A few studies have confirmed that environmental variation causes selection for different traits in AMF. Such selection can lead to differences in symbiotic functioning among AMF isolates according to the contrasting climates or soil conditions in the areas of their origin (Mena-Violante et al., 2006; Antunes et al., 2011; Sochacki et al., 2013). For example, even short-term waterlogging can change the AMF community present in the roots of the host plant (Yang et al., 2016) and the fine root endophyte (*Glomus tenue* (Greenall) I.R. Hall) seems to be particularly well adapted to this stress (Orchard et al., 2016). Moreover, local adaptation to varying environmental conditions may produce more important variation in AMF than the basic functional differences between coexisting taxa (Sanders, 2002). In a recent study, Millar and Bennett (2016) found that abiotic stresses can have effects on AMF fungi that are independent of those on the host plant and can change the composition of AMF communities. Such an outcome could mean the loss of important individual species. Furthermore, depending on phylogenetic relatedness, the functioning of the AMF or plant community could have strong negative ecological consequences (Maherali and Klironomos, 2007). This change in the composition of the AMF community could make a plant dependency on an AMF symbiosis (or a particular fungal species) less suitable. Therefore, changes in the composition of the AMF community could alter feedback to the plant community and beyond (Millar and Bennett, 2016). However, the potential for AMF to adapt to new conditions can be a particularly important feature of an organism with limited dispersal capability.

### 4.3.3 Protection Against Biotic Stresses

It is widely accepted that AM symbioses can be effective in reducing the impact on host plants from soilborne pathogens, since AM plants have been observed to receive protection from pathogens relative to their nonmycorrhizal counterparts in experimental studies (Newsham et al., 1995; Filion et al., 1999, 2003; Borowicz, 2001; Carlsen et al., 2008). Many studies have reported that AMF can protect plants against a wide range of plant pathogens, including soilborne fungal and bacterial pathogens, nematodes, phytopathogenic insects, and parasitic plants (Whipps, 2004; Pozo and Azcón-Aguilar, 2007). In contrast, mycorrhizal plants are known to develop greater susceptibility to viruses (Whipps, 2004). However, a more recent study indicated that colonization by *F. mosseae* appears to exert a beneficial effect on tomato plants in attenuating the disease caused by tomato yellow leaf curl Sardinia virus (Maffei et al., 2014).

*Investigation of the protection against abiotic and biotic stresses has indicated that there is functional diversity but the extent of our knowledge is somewhat limited. An important factor is the interaction between AMF and other microbes, which has scarcely been tested with respect to its contribution to functional diversity.*

The AM symbiosis is characterized as providing *broad spectrum* protection against a wide range of organisms that attack plants. Although most of the studies have tested the role of a single or a very limited number of species of AMF in protection against plant pathogens, most of the positive effects were observed with isolates belonging to the Glomeaceae. This supports evidence that intraradical colonization by members of the Glomeraceae is correlated with effective protection against plant root infection by common soil pathogens. The results suggest that the functional trait, protection against soil pathogens, is conserved within this family and may have evolved to help the plants overcome the harmful effects caused by particular pathogens (Maherali and Klironomos, 2007). However, studies comparing different fungal species or isolates belonging to the genus *Glomus* have identified that functional diversity of AMF as biocontrol agents against *Verticillium* (Kobra et al., 2009) or nematodes (Habte et al., 1999) can be found at the isolate level. Acquisition of nutrients by plants has been suggested as being involved in the bioprotection conferred by AMF. Therefore, the use of isolates belonging to the Gigasporaceae family, with their greater hyphal density in soil, would functionally complement the Glomeraceae family in bioprotection by enhancing mineral nutrition of host plants (Maherali and Klironomos, 2007).

Further functional diversity in bioprotection can result from the interaction with other soil microbes. Root colonization by AMF results in changes in the microbial community in the mycorrhizosphere, which can be positively influenced by certain mycorrhiza helper bacteria (MHB) strains (Lioussanne et al., 2010). The MHB can be found in the cytoplasm of AMF spores (Lumini et al., 2007), colonize the surface of AMF spores and hyphae (Xavier and Germida, 2003), and may also be located inside the spore walls (Walley and Germida, 1995). Characterization of the bacteria isolated from AMF spores and mycorrhizal cultures showed that they have antifungal activity against particular pathogens and can solubilize phosphate (Cruz et al., 2008; Cruz and Ishii, 2011). There are several examples of the antagonistic effects on pathogens resulting from the combination of AMF with other beneficial microorganisms, such as plant growth-promoting rhizobacteria (PGPR) (Järderlund et al., 2008; Akhtar and Siddiqui, 2008; Singh et al., 2013). Therefore, experiments focused on the effects of a single microbial species may overlook important multipartite interactions of more naturalistic microbial communities and underestimate the role of other beneficial organisms associated to AMF.

### 4.3.4 Improvement in Soil Structure

In addition to enhancing the volume of soil able to be exploited (Bedini et al., 2009; Caravaca et al., 2006), the ERM can enmesh soil particles improving soil structure and stability and therefore water retention, by the secretion of exudates including glomalin and glomalin related soil proteins (GRSP) (Andrade et al., 1998; Rillig, 2004b; Rillig and Mummey, 2006; Hallett et al., 2009). However, the separate contributions from these different mechanisms to the benefits from AMF are not always evident and clearly considerable functional diversity exists between these fungi in relation to aggregation and aggregate stability (Miller and Jastrow, 2000). Schreiner et al. (1997) reported a greater improvement of aggregate stability promoted by *G. mosseae* relative to that achieved by *G. etunicatum* or *Gi. rosea*, which is not consistent with the usually reported greater ERM development for members of Gigasporaceae in general. Although Bedini et al. (2009) did not find a significant relationship between GRSP and ERM data, such as total hyphal length and hyphal density, they observed that ERM parameters (hyphal length and hyphal density) for the AMF tested were correlated with mean weight diameter (MWD) of 1–2 mm diameter macroaggregates. These aggregates are considered to be sensitive to short-term treatments of soil. These same authors also reported that two isolates of *G. mosseae* (IMA 1 and AZ225C) produced significantly different amounts of total GRSP (Bedini et al., 2009). Using less discriminative methods, an earlier study also showed differences in glomalin production by different species of *Glomus* and *Gigapspora* (Wright et al., 1996). However, Bedini et al. (2009) found no effect of a *G. mosseae* isolate on aggregate MWD. In contrast, there were interspecific differences between *G. mosseae* and *G. intraradices* and interisolate differences between two isolates of *G. intraradices* (IMA 5 and IMA 6) in the MWD of aggregates present.

## 4.4 AMF DIVERSITY ASSOCIATED WITH THE MANAGEMENT OF DIFFERENT ECOSYSTEMS

*There is evidence that farming practices that aim to maintain AMF can also contribute to increased diversity.*

The extent to which different soil management practices affect soil microbial diversity has been an important topic for investigation for some time. Studies focused on this issue have shown that soil microbial diversity not only varies with soil management but so too does soil functioning (Reeve et al., 2010; Avio et al., 2013). In addition, research on the physiology and ecology of AMF has also contributed to a greater understanding of the multiple roles that mycorrhiza can perform in an agroecosystem and how agricultural management practices influence the efficacy of the AM symbiosis. For example, there is considerable diversity within the Glomeromycota, particularly in undisturbed soils, such as those under grassland and forest (Zangaro et al., 2013). Furthermore, diversity is affected by various factors, including host plant, soil, environmental conditions, and cropping practices (Hayman, 1982; Wang and Li, 2013).

An increasing number of studies have shown that intensive land use threatens soil biodiversity (e.g., Verbruggen et al., 2010; Tsiafouli et al., 2015), where intensive arable systems commonly have a lower level of soil biodiversity than grassland (Tsiafouli et al., 2015) or organic farms (Tuck et al., 2014). Land use intensification may not only reduce biodiversity in the soil but can also contribute to several environmental problems, such as the eutrophication of surface water, reduced aboveground biodiversity, and climate change (Foley et al., 2005). It has been suggested that a diverse AMF population could improve plant fitness and productivity or even exhibit greater potential to protect host plants against pathogens than would a single AM fungal species. Such a proposal assumes that larger communities have a greater probability of containing greater phenotypic trait diversity and complementarity. Therefore, the loss of a small number of species or functional groups in intensive land use systems would more easily impair ecosystem functions than would be the case in natural ecosystems (Hunt and Wall, 2002).

Different agricultural practices can exert an enormous selective pressure on AMF, shaping their community structure and evolution by modifying several of their biological features, such as strategy for sporulation, resource allocation, and spatial distribution (Verbruggen and Kiers, 2010). In addition, AMF seem to be highly sensitive to anthropogenic activities (Giovannetti and Gianinazzi-Pearson, 1994). For example, numerous studies have shown that arable soils have a smaller AMF species richness in comparison to natural woodlands and grasslands, indicating that AMF diversity, effectiveness, abundance, and biodiversity decline in agroecosystems subjected to high-input practices (Lumini et al., 2010; Borriello et al., 2012). Agricultural soils can differ in AMF species richness and composition because management practices differ significantly, as in the case of low- and high-input agricultural systems (Ryan et al., 2000) (see Section 2.2). Intensive agricultural practices, such as monocropping or crop rotations that include nonmycotrophic crop species, the use of bare fallow periods, excessive fertilizer application, control of pest and diseases using fungicides, and the impact of tillage on the disruption of hyphal networks can all contribute to a reduction in AMF population biodiversity (Helgason et al., 1998; Daniell et al., 2001, Merryweather 2001; Jansa et al., 2002).

Tillage regimes have been considered to favor species that can colonize through fragmented hyphal networks and infected root pieces (Biermann and Linderman, 1983; Hamel, 1996). Therefore, as isolates belonging to the Gigasporaceae family tend not to colonize hosts from infected root pieces or ERM fragments, it explains the reduced number of species of this family identified in fields under inversion tillage (Jansa et al., 2003; Castillo et al., 2006a). Since isolates from Gigasporaceae have been shown to complement other AMF families functionally (Maherali and Klironomos, 2007), their exclusion from agricultural systems represents a potential loss of useful functional diversity for crop hosts. On the other hand, Glomeraceae isolates are considered a cosmopolitan AMF family, since it is possible to observe species spread across global agricultural systems (Rosendahl et al., 2009). As anastomosis of hypha occurs with individual AMF of the same isolate within this family, it is possible that the ability to rapidly form interconnecting mycelium networks could be significant for the survival and maintenance of these fungi in soils subject to regular tillage (de la Prividencia et al., 2005; Rosendahl and Matzen, 2008).

The excessive use of phosphorus fertilizer can inhibit mycorrhizal colonization and growth as well as reduce AMF community richness, which results in declining diversity (Egerton-Warburton et al., 2007; Alguacil et al., 2010; Borriello et al., 2012; Liu et al., 2012b). However, fertilizers application has a differential effect on AMF species in the soil. The effect of phosphorus application in the AMF diversity seems to be related with the ability of different species to colonize the roots of the host plant at different levels of P in the soil. Sylvia and Schenck (1983) reported that *Gi. margarita*, *G. clarum*, and *Glomus mossae* were able to colonize the roots of *Paspalum notatum* Flugge, even at very large concentrations of P in the soil, and the spore concentration of these species increases in the soil due superphosphate application. In contrast, sporulation by *G. etunicatum*, *G. macrocarpum*, and *G. gigantea* was reduced. Bhadalung et al. (2005), studying the effect of long-term liberal supply of P and N to the soil, found that *Acaulospora* sp. 1 was insensitive to fertilizer application, *Entrophospora schenckii*, *G. mosseae*, *Glomus* sp. 1, *G. geosporum*-like, and *Scutellospora fulgida*, slightly sensitive, while three unidentified species of *Glomus* were found to be very sensitive to the application of fertilizer.

Crop rotation with nonmycotrophic plants results in a decrease of the AMF activity, whereas the use of highly mycorrhizal host plants increase AMF inoculum potential of the soil and colonization of the subsequent crops (Black and Tinker, 1979; Gavito and Miller, 1998; Karasawa et al., 2002). Nevertheless, Gavito and Miller (1998) suggested that AMF populations can be built up and the inhibitory effects of a nonmycotrophic crop can be reversed by sowing a mycorrhizal crop. Mycotrophic weeds can play the same role. For example, Kabir and Koide (2000) showed that the mycorrhizal weed, dandelion (*Taraxacum officinale* Weber ex Wigg.), which is a typical weed of arable soil, was able to enhance colonization of maize relative to that following clean fallow (Fig. 4.1) over a period of 8 weeks.

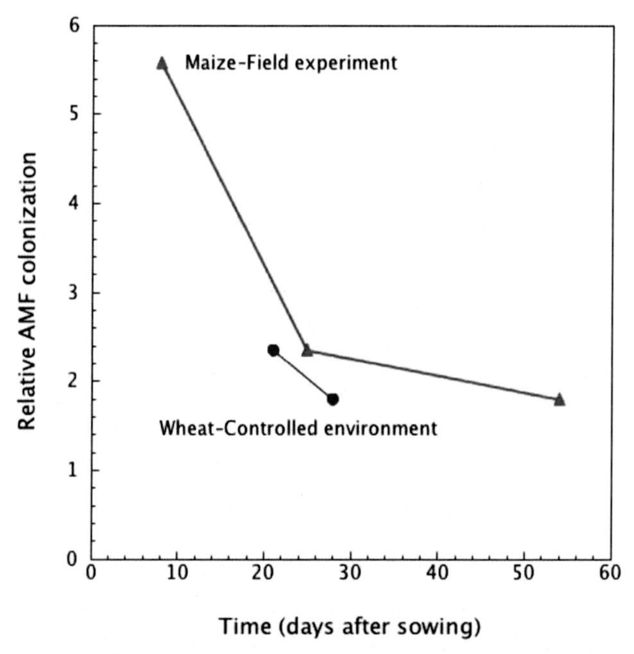


**FIGURE 4.1** The proportion of root colonized by arbuscular mycorrhizal fungi (AMF) from Developer plants relative to that of the same cereal plant dependant mainly on spores for colonization. Results for maize (red line) based on Kabir and Koide (2000). Results for wheat (blue line) based on Brito et al. (2013).

Between 3 and 4 weeks after sowing, results for wheat following grass weeds were similar to those for maize following dandelion (Fig. 4.1) (Brito et al., 2013).

The increasing demand for sustainable agricultural production has resulted in greater interest in the use of beneficial soil microorganisms and the need for their manipulation. Management of native populations of AMF is recognized as a sustainable strategy for agriculture because it can optimize the use of chemicals and energy in agriculture and minimizing environmental degradation (Jeffries and Barea, 2012). Such biological interventions are made more attractive as the use of chemicals for fumigation and disease control has been progressively discouraged and the long-term availability of phosphatic fertilizers is now measured in decades (Atkinson, 2009). Furthermore, the difficulties associated with AMF inoculant production and application, discussed in Box 3.1, combine to encourage the investigation of how to ensure sustainable exploitation of the potential benefits associated with indigenous AMF.

It is clear that soil management practices, which minimize negative effects on soil biota but provide desirable agricultural benefits, need to be further developed. The enhancement of AMF abundance and biodiversity within cropping systems requires an integrated approach using different production techniques (see Chapter 2).

## 4.5 CONCLUSIONS

Despite the fact that AMF are the most abundant symbionts in ecosystems, only about 240 species, based on AMF spores morphological features, have been described. The use of molecular techniques in ecological studies of AMF have revealed a much larger number of taxonomic groups, indicating that the overall AMF diversity is significantly underestimated, but does not differentiate between functional and biological variation in AMF impacts on plant growth and community composition.

The evidence of high functional diversity in AMF from genetically different AMF isolates, even within species, may play an important role in determining plant diversity, productivity, and ecosystem variability. However, in spite of the evidence of the involvement of the exchange of phosphorus and photosynthate between plants and AMF, colonization rates and growth of ERM, other mechanisms behind the functional diversity and the extent to which these functions are distributed within and among different taxa are still unclear.

Maintenance of a sustainable mixed plant population depends on the preservation of a diverse AMF population, based on the concept that larger AMF communities have greater probability of containing a greater diversity of phenotypic traits. Such an approach may ensure the essential number of species required for ecosystems functioning. Moreover, the evidence of coexistence of distinct evolutionary lineages, due to the phylogenetic trait conservatism, can enhance the ecosystem function because of functional complementarity among those same lineages. Thus any change in their populations or loss in diversity may result in changes of agro-ecosystem productivity or loss of resilience against adverse conditions due to shifts in the AMF abilities to sustain multiple functions in above- and belowground ecosystems.

Since it was demonstrated that management practices such as crop rotation, tillage, and phosphorus fertilization influence AMF diversity and thus AMF functioning, soil management practices that minimize negative effects on AMF diversity will enable the proper functioning of ecosystems and will provide the desired agricultural benefits. However, such practices need to be further developed. The combination of crop sequence selection and tillage practices that maintain the ERM intact appears to be a possible approach to selecting a consortium of AMF that can be induced to colonize and improve the productivity of more than one crop.

AMF diversity is critical to ensure biodiversity and ecosystem functioning, but ensuring the presence of a diverse AMF population in soils remains a challenge. However, strategies involving soil ecological engineering, using cropping sequences associated with appropriate tillage and cultivation practices to maximize the contribution of soil biological processes to sustainable ecosystem functioning need to be enhanced if food security is to be guaranteed at the same time as minimizing negative environmental impacts.

Chapter 5

# Impacts on Host Plants of Interactions Between AMF and Other Soil Organisms in the Rhizosphere*

## Chapter Outline



The last 30 years has seen a huge increase in the detailed understanding of the microbial environment surrounding plant root systems and of the processes involved in the establishment of the mycorrhiza symbiosis (Martinez-Garcia et al., 2013). Knowledge of the interactions between microbes, particularly how they communicate with each other and with higher plants, has also expanded dramatically (Gobbato, 2015). This development has also allowed a more holistic approach to the investigation of mycorrhiza and the possibility for optimizing the beneficial aspects of the symbiosis.

*Arbuscular mycorrhizal fungi (AMF) are an important and ancient component of soil microbial biomass. They inevitably interact with a broad range of soil microbes, not least because there is a concentration of biota in the rhizosphere.*

*With Luís Alho and Sabaruddin Kadir.


**Functional Diversity of Mycorrhiza and Sustainable Agriculture.**
**DOI: http://dx.doi.org/10.1016/B978-0-12-804244-1.00005-8**

## 5.1 INTERACTIONS BETWEEN AMF AND OTHER SOIL MICROBES

The recycling of plant nutrients and carbon in the soil is dependent on the activity of bacteria and fungi, including arbuscular mycorrhizal fungi (AMF). Over the 460-million-years of its existence (Selosse et al., 2015), the symbiosis between two-thirds of the plants on earth and AMF is believed to have developed from the simple exchange of carbon for P and diversified to the extent that the host plant may receive a number of mineral nutrients, mostly but not exclusively those considered poorly mobile in soil, protection from toxic metals or metalloids, defense against pathogens, resistance against drought, and enhancement of the structure of the soil in which it grows. In providing these services, AMF inevitably interact with the millions of microbes also occupying the volume of soil they and their host plants share. For example, the number of bacteria in soil is typically reported as $10^9$ $g^{-1}$ (Table 5.1). However, the distribution of microbes in the soil is not uniform but is several times greater in the *rhizosphere* – the soil volume immediately around the root system of a plant – than in the bulk soil. The ratio can range from 28 to 53 for bacteria, 1.3 to 12 for fungi (Bowen and Rovira, 1999). Nor is the rhizosphere restricted to fungi and bacteria, as these can also be the subject of attack by herbivores and predators among the protozoa and nematodes (Table 5.1), which also show an increased presence of ~10-fold in rhizosphere soil (Bowen and Rovira, 1999). Although the main concentration of microbes is in the topsoil, the upper 20–25 cm of the profile, bacterial cells at least are active in deeper horizons (Blume et al., 2002) and between 35% and in excess of 58% of the total microbial biomass can be present below 25 cm (Fierer et al., 2003; Schütz et al., 2010). A significant factor in the *rhizosphere effect* on the distribution of microbes in the soil is the release from growing roots of a broad spectrum of carbon compounds that can support their energy requirements. In addition, there are specific compounds released that are important in signaling between microbes and the plants via the root system. But these compounds are also involved in the communication between different groups of microbes. There is increasing evidence that in the rhizosphere and in the soil volume, which is associated with mycorrhiza and the AMF extraradical mycelium (ERM) – the *mycorrhizosphere*, the bacterial community is very much determined by the presence of the fungal component (de Boer et al., 2005, 2015; Bonfante and Anca, 2009). All of these aspects are important in understanding the effects on mycorrhizal host plants of the interactions between AMF and other soil microbes.

*A key AMF interaction is with a special group of bacteria (rhizobia) that are capable of taking nitrogen from the atmosphere and converting it to ammonia. Both form a symbiosis with legume plants.*

**TABLE 5.1** The concentration of main groups of microflora and fauna in soil

| Organisms | | Number | Units | References |
|---|---|---|---|---|
| **Microflora** | | | | |
| Archaea | | $4.4-2.4 \times 10^4$ | amoA gene copies[a] $g^{-1} \times 10^3$ | Bates et al. (2011); Nicol et al. (2008); Habteselassie et al. (2013) |
| Bacteria[b] | Tropical soil with large organic matter content | $0.4-2.0 \times 10^6$ | cfu $g^{-1} \times 10^3$ | Vieira and Nahas (2005) |
| | Cropland | $3.91-5.69 \times 10^5$ | $10^3$cells $g^{-1}$ | Bressan et al. (2015) |
| | Grassland | $6.69 \times 10^5$ | $10^3$cells $g^{-1}$ | Bressan et al. (2015) |
| Actinomycetes | | $0.7-2.9 \times 10^3$ | cfu $g^{-1} \times 10^3$ | Ghorbani-Nasrabadi et al. (2013) |
| | Tropical soil with large organic matter content | $7.96-8.80 \times 10^3$ | cfu $g^{-1} \times 10^3$ | Vieira and Nahas (2005) |
| Total fungi | | $5-8 \times 10^2$ | cfu $g^{-1} \times 10^3$ | Vieira and Nahas (2005) |
| Cyanobacteria | | $0.32-8.2 \times 10^5$ | cfu $g^{-1} \times 10^3$ | Hunt et al. (1979)[c] |
| Algae | | $1.0-1.3 \times 10^4$ | cfu $g^{-1} \times 10^3$ | Hunt et al. (1979)[c] |
| **Microfauna** | | | | |
| Total protozoa | Pasture | 29.20 | $g^{-1} \times 10^3$ | Griffiths et al. (2000) |
| | Grassland | 1.13 | $g^{-1} \times 10^3$ | Griffiths and Ritz (1988) |
| | Arable soil | 5.55 | $g^{-1} \times 10^3$ | |
| Flagellates | Annual grass | 0.50 | $g^{-1} \times 10^3$ | Hungate et al. (2000) |
| | Arable soil | 0.57–13.95 | $g^{-1} \times 10^3$ | Darbyshire and Greaves (1967) |

(*Continued*)

**TABLE 5.1** (Continued)

| Organisms | | Number | Units | References |
|---|---|---|---|---|
| Ciliates | Annual grass | 0.01 | $g^{-1} \times 10^3$ | Hungate et al. (2000) |
| | Arable soil | 0.03–0.28 | $g^{-1} \times 10^3$ | Darbyshire and Greaves (1967) |
| Amoebae | Cropped soil | 2.33–51.98 | $g^{-1} \times 10^3$ | Darbyshire and Greaves (1967) |
| Testacae | Pasture | 0.6 | $g^{-1} \times 10^3$ | Bamforth (1971) |
| Nematodes | Grassland | | | |
| | Total | 15.18 | $g^{-1} \times 10^3$ | Sohlenius and Sandor (1987) |
| | Plant feeder | 6.32 | $g^{-1} \times 10^3$ | |
| | Fungal feeder | 2.56 | $g^{-1} \times 10^3$ | |
| | Bacterial feeders | 3.88 | $g^{-1} \times 10^3$ | |
| | Omnivores | 1.10 | $g^{-1} \times 10^3$ | |
| | Barley Field | | | |
| | Total | 9.92 | $g^{-1} \times 10^3$ | |
| | Plant feeder | 2.34 | $g^{-1} \times 10^3$ | |
| | Fungal feeder | 2.56 | $g^{-1} \times 10^3$ | |
| | Bacterial feeders | 4.80 | $g^{-1} \times 10^3$ | |
| | Omnivores | 0.22 | $g^{-1} \times 10^3$ | |

[a] *Assumes there are between 1 and 3 amoA (ammonia monooxygenase) gene copies per archaeal cell.*
[b] *Assumes that less than 0.1%–10% of total bacteria are culturable.*
[c] *Counts of autotrophic microbes in the soil's crust (1 cm depth samples).*

### 5.1.1 The Tripartite Interaction Between AMF, Rhizobia, and Legumes

The interaction between microbes, including AMF, is not confined to the bulk soil and the rhizosphere but potentially can also take place within plant roots. As well as forming mycorrhizal symbioses with AMF, a number of plants also establish a symbiosis with nitrogen fixing bacteria. The symbiosis of major significance in productive ecosystems is that of members of the legume family, the Fabaceae and trees of the genus *Parasponia* with rhizobia (Vessey et al., 2004). "Rhizobia" is the common collective name for several symbiotic bacteria genera, including *Azorhizobium*, *Bradyrhizobium*, *Mesorhizobium*, *Rhizobium*, and *Sinorhizobium*. The symbionts combine to form novel structures, nitrogen-fixing nodules, in plant roots that are capable of fixing nitrogen gas from the atmosphere. The symbiosis is estimated to contribute annually some 40 million tonnes of nitrogen from the atmosphere to agricultural production systems (Herridge et al., 2008). In comparison to the symbiosis with AMF, it developed relatively recently, only being present for 60 million years in the fossil record.

#### *Colonization of Roots by Mycorrhizal Fungi*

The process of the colonization of a root by an AMF does not start with the formation of a hyphopodium (See Section 6.2) but with stimulation of the fungus by components in root exudates (Harrison, 2005). The specific compounds are strigolactones (Akiyama et al., 2005). These chemical signals can stimulate the germination of spores, switch on the genes responsible for the signaling system of the AMF as well as affect activity in the organelles (mitochondria) responsible for generating energy (adenosine triphosphate, ATP) and reducing power (reduced nicotinamide adenine dinucleotide, NADH) in the AMF (Besserer et al., 2006). This allows the AMF to produces diffusible signal compounds, Myc Factors, lipochitooligosaccharides (LCOs), that are released into the soil and which in turn affect changes in the structure and physiology of the host root (Maillet et al., 2011). Recognition of the Myc Factors likely requires two receptor molecules located on the plasma membrane of root epidermal cells (Paszkowski et al., 2006; Roberts et al., 2013). These compounds can stimulate root branching, affect root hair formation and growth, as well as initiate major changes in the cells of the epidermis close to the AMF source, including the formation of a prepenetration apparatus (PPA) used to guide the fungal hypha from the hyphopodium on the root surface (the site of fungal contact – SFC) through the epidermal cell and into the inner cortex of the root (Box 5.1). In the inner cortex the hypha emerges from the tube-like PPA of the epidermal cell, where it grows and branches in the intercellular spaces. These hyphae then induce a PPA-like structure in cells of the inner cortex, which they enter, branch and form arbuscules, the essential site for nutrient exchange between fungus and host.

> *There is a great similarity in the processes of colonization by AMF and rhizobia, particularly in the communication between microbe and plant.*

## *Colonization of Roots by Rhizobia*

For this symbiosis, development starts with release from the roots of the host plant of signal molecules, mainly flavonoids (Box 5.1) but also included are simple sugars, amino acids, dicarboxylic acid, and hydroxyaromatic acid. These signals not only act to attract rhizobia to the roots of potential hosts but also stimulate changes in the bacteria that activate the rhizobial genes required for initiating the symbiosis and the eventual production of nodules. A critical consequence of the activation of these genes is the production and release into the soil of LCOs, the so-called nodulation factors (Nod Factors). For the legume to participate in the symbiosis, the necessary expression of host genes – the nodulin (nod) genes – have to be activated and this is the role of the Nod Factors. Perception of Nod Factors in the roots of a host initiates numerous changes in the plant, the details of which vary depending on the host. However, these changes can include the curling of root hairs and the formation of subcellular structures and meristems that form the nodules to house the bacteria (Box 5.1). The curling and branching of root hairs entraps rhizobia that may be attached to the hair surface. Rhizobia penetrate the root hair cell wall following localized hydrolysis of the wall (Callaham and Torrey, 1981) and enter an *infection thread*, which is formed by invagination of the plasma membrane (plasmalemma) and generation of new wall material in the form of a tubular lining. The rhizobia therefore remain extracellular as the infection thread extends along the root hair and across the lumen of the cell and then into the root cortex. There, in the mid-cortex, the rhizobia leave the infection thread and enter cells, which have also been undergoing genetically controlled changes induced by the Nod Factors to establish a *nodule primordium*. Within these cells the rhizobia are enveloped by membranes formed by the host, change their shape to become *bacteroids* and within these organelle-like structure, called *symbiosomes*, begin fixing nitrogen. Symbiosomes can be considered as organelles, similar to mitochondria or chloroplasts, being surrounded by a specialized plant membrane that permits metabolite exchange. The Nod Factors form an essential part of the signaling system that controls all stages of the infection process, including the growth of the infection thread, initiation of nodule formation, and the transformation of the bacteria into bacteroids (Ovchinnikova et al., 2011). Mature legume nodules may be *indeterminate*, maintaining the apical meristem and growing by cell division and expansion, or the meristem development may be transient, so that nodules are *determinate*, only growing by cell enlargement (Box 5.1). In general, temperate legumes, such as pea and alfalfa, develop indeterminate nodules, which are initiated in the pericycle and inner cortex, and grow between the root cells and emerge as club-shaped organs external to the root axis. In contrast the determinate nodules formed by tropical legumes, such as soybean (*Glycine max* L. Merr.), are large drop-shaped structures that are initiated in the outer cortex of the roots (Hirsch, 1992).

**BOX 5.1 Cellular and Physiological Changes During Initial Colonization of Roots by Arbuscular Mycorrhizal Fungi (AMF) and Rhizobia**

Most members of the Fabaceae form symbiotic relationships with both AMF and nitrogen-fixing bacteria, the latter being commonly referred to as rhizobia. The annual barrel medic (*Medicago truncatula* Gaertn.) has become particularly well studied as a model plant for understanding the physiology of leguminous plants, including the colonization processes by these two groups of microbes (Young et al., 2011). Once an AM fungus establishes signal exchange with the root of the *M. truncatula* host, a hyphopodium forms on the root surface. The epidermal cell below begins to undergo major internal reorganization (Reinhardt, 2007). It starts with the cell nucleus migrating to a position below the hyphopodium and then moving across the cell lumen toward the inner periclinal wall opposite the SFC (Genre et al., 2005). It leaves behind, in its position under the hyphopodium, a collection of microtubules and microfibrils of the cytoskeleton together with cisternae of endoplasmic reticulum (ER). These structures become organized into a finger-like cytoplasmic column, which tracks the route of the nucleus across the cell lumen. A large number of microtubules and bundles of actin microfilaments become aligned parallel to the column and the very dense mass of ER cisternae, which is in reality a hollow tube joining the nucleus to the SFC. The whole arrangement linked to the nucleus is recognized as the PPA. It appears that an invagination of the plasma membrane takes place to line the hollow tube. Once the cytoplasmic column of the PPA has completed the crossing of the cell lumen, the nucleus migrates to the side and a fungal hypha, formed as an outgrowth of the hyphopodium, penetrates the cell wall, possibly by a local degradation caused by the release of enzymes coupled with some mechanical force (Harrison, 1999a). The hypha grows down the hollow tube (Genre et al., 2005) and does not penetrate into the cytoplasm but remains within the apoplast. This general process may be repeated as the fungus penetrates the outer layer of the root cortex or this may take place via the intercellular spaces. Similarly, penetration of the epidermis may be through the wall separating two cells but in this case penetration to the inner layer of the cortex is intercellar. It appears that at least one stage of infection has to be intracellular for colonization of the inner cortex to be successful (Genre et al., 2008).

The process of forming an arbuscule within an inner cortical cell is somewhat similar to the initial stage of intracellular penetration by a hypha. Where a

(*Continued*)

**BOX 5.1 (Continued)**

hyphal tip makes contact with a cortical cell wall, a localized concentration of ER develops within the cell. This is also associated with the cell nucleus becoming enlarged and moving to the center of the cell (Genre et al., 2008). Invagination of the plasma membrane takes place and the hypha penetrates into the cell, where it undergoes dichotomous branching to form several orders of branches and fill much of the cell. The host cell vacuole may become partly fragmented, or appear so because of distortions of the tonoplast (Pumplin and Harrison, 2009), and ER, large numbers of plastids, and mitochondria congregate around the branches (Hause and Fester, 2005). Actin microfibrils and microtubules form a complex cytoskeleton throughout the cell (Genre et al., 2008). The invaginated plasma membrane of the host cell, often termed the peri-arbuscular membrane, is significantly modified. It contains phosphate transporters and shows intense ATPase activity, especially where it surrounds finer branches of the arbuscule structure (Harrison, 1999b; Hause and Fester, 2005). Between the peri-arbuscular membrane and the plasma membrane of the fungus is cell wall material of the host, but the structure is not consolidated into a secondary wall and the space between the two membranes has an acidic pH, concomitant with the transfer of nutrients between the partners (Rich et al., 2014).

For the establishment of its symbiosis with the rhizobia *Sinorhizobium meliloti, M. truncatula,* first has to release the flavone 7,4′-dihydroxyflavone (Dhf) (Zhang et al., 2009). This flavonoid binds directly with receptor proteins, the product of the NodD gene (the only bacterial gene involved in nodulation that is permanently active) within *S. meliloti* and thus activates the bacterial genes required for nodulation of the host plant (Nap and Bisseling, 1990). The various species of legumes release different combinations of flavonoid signals, and it is considered that specificity in the binding with the receptor proteins of the bacteria is one factor in the selectivity of the symbiosis. Mixtures of flavonoids can be more effective in the establishment of root nodules than single compounds in that they encourage the activity of some rhizobia but can be antagonistic to others (Cooper, 2007). The nodulation genes of the rhizobia are required for synthesis of Nod Factors, the compounds that need to be perceived by the host plant to initiate the next steps in the formation of nodules. Typically the exudates from the host roots attract the symbiotic bacteria in the rhizosphere to the root surface, and some attach to the root hairs. This appears to be a two-stage process, with end-on initial attachment to a receptor protein followed by a structural linkage involving either cellulose fibrils (Smit et al., 1987) or proteinaceous, fibrillar structures – fimbriae – (Vesper and Bauer, 1986). These structures, formed by the microbe, allow bacterial cells to aggregate at the location following division so that a colony develops at each point of attachment. Collectively these colonies are referred to as infection foci. The host plant detects the bacterial Nod Factor return signals from *S. meliloti,* in the form of LCO molecules (Long, 1996), through receptors, likely lectins, distributed on the plasma membrane localized at the tips of growing root hairs (Dazzo et al., 1978; Law and Strijdom, 1984; Roberts et al., 2013).

(*Continued*)

**BOX 5.1 (Continued)**

Nod Factors stimulate a number of different reactions in root hair cells of the host legume, including an initial decrease in osmotic potential, possibly resulting from calcium uptake; the modification of growth; depolarization of their plasma membrane; rapid fluctuations in the levels of intracellular free calcium (called calcium spiking); modifications to the cytoskeleton and stimulate the formation of the *preinfection thread* in deformed root hairs (van Brussel et al., 1992). They also stimulate cortical cell division at the sites of nodule primordia formation; inhibit the system that generates reactive oxygen; acting together with endogenous flavonoids in the root they perturb auxin flow in roots and induce the activation of regulatory plant genes involved in nodule formation (*nodulin genes*).

In epidermal cells with an emerging root hair, cytoplasm, including the spherical nucleus, is concentrated in the subapical region of the developing protrusion, with a vesicle-rich zone at the tip. The region of dense cytoplasm, the organization of which is maintained by bundles of actin filaments, contains ER, mitochondria, plastids, and Golgi bodies. The main part of the cell contains a large vacuole. Cortical microtubules are oriented obliquely (at varying angles) or transverse to the long axis of the root, especially around the location of the emerging hair, where they are also parallel to one another. The nucleus tracks the polar growth of the extending tip of the hair but remains at a distance from it. As the root hair growth declines, the microtubular cytoskeleton becomes progressively helical and the nucleus changes to ellipsoid. The vacuole progressively extends into the hair and the nucleus finally moves to lie against the cell wall in the lower part of the root hair. Colonized root hairs undergo major changes in the pattern of growth. These changes can result in curling or branching and lead to the entrapment of colonies of the bacteria within an infection pocket, either formed by the curling of the tip back on itself, rather like a shepherds crook or by a newly established branch growing toward the established part of the hair (Oldroyd and Downie, 2004). From this pocket, preinfection threads then form as invaginations of the cell plasma membrane over which newly synthesized cell wall material is deposited. A network of endoplasmic microtubules, which formed a network around the nucleus, progressively replaces the existing helical arrangement of cortical microtubules. A new network of cortical microtubules forms parallel to the axis of the root hair. Then the nucleus migrates to the tip of the root hair, and during this time the microtubular cytoskeleton gradually concentrates in the region between the nucleus and the root hair tip.

Actual infection by *S. meliloti* starts with a very localized hydrolysis of the root hair cell wall, a process involving the alteration and degradation of cell wall polysaccharides. Microtubules are recruited for the formation of the infection thread and accumulated to form dense parallel arrays extending from the infection pocket. The microtubule cytoskeleton transforms into a dense network surrounding the extending infection thread and connects the nucleus to the infection thread tip. Longitudinal microtubules form parallel to and in close contact with the infection thread.

(*Continued*)

**BOX 5.1 (Continued)**

In addition to the Nod Factors, the successful invasion of *M. truncatula* requires the release of an acidic exopolysaccharide, called succinoglycan (Jones and Walker, 2008). This compound acts as a signal to the host plant to permit the entry of *S. meliloti* into the preinfection threads. A second exopolysaccharide, galactoglucan, has also been identified that has similar effects to succinoglucan but its formation seems to be induced when inorganic phosphate levels in the soil are very low (Krol and Becker, 2004; Glenn et al., 2007) and there are sufficient numbers of *S. meliloti* present (Pollock et al., 2002).

Once entry has taken place, rhizobia proliferate in what has become the *infection thread* as it develops along the root hair, so they maintain a position close to the leading end of the tube. The thread grows across the lumen of the cell and then invades cells of the cortex. Similar changes in the arrangement of the cytoskeleton take place as infection threads grow from the activated root hair cells to the first cell layer of the outer cortex. As the infection thread approaches the next inner cell, the latter forms a preinfection thread by establishing a cytoplasmic bridge and its nucleus migrates toward the point of transfer. Localized disruption of the cell wall takes place, microtubules accumulated at both sides of the transfer location, and the nucleus of this cell becomes attached to the infection thread, which follows along the cytoplasmic bridge toward the next cell.

Nodulation requires the coordination of the initial epidermal infection by rhizobia with cell divisions in the underlying cortex. Even before the infection thread has crossed the epidermis, pericycle, and cortical cells in a zone opposite a protoxylem pole respond in a local manner to the rhizobia. In pericycle cells this is reflected by the rapid induction of a nodulin gene and by rearrangements of the cytoskeleton to one characterized by endoplasmic microtubules (Yang et al., 1993; Timmers et al., 1999). These cells undergo a limited number of anticlinal and periclinal divisions to form a localized bilayer pericycle (Timmers et al., 1999). Next, Nod Factors induce cells of the inner cortex to divide, although rarely those of the endodermis, and form the initial nodule primordium. Prior to division the nucleus swells and moves from the periphery to the center of the cell, remaining linked to parietal cytoplasm by cytoplasmic strands that cross the central vacuole. Mitosis in cells near the middle of the root cortex and next to the initial primordium, results in the generation of the nodule meristem. This begins with each cell undergoing multiple divisions to create groups of meristematic cells that aggregate into a division center. The nodule meristem may continue to produce new cells that can become infected or, once the initial period of cell division is complete, no further divisions occur. In both cases cells expand as they become packed with bacteria. Cells of the outer cortex undergo the same initial structural changes as those of the inner cortex in terms of nucleus size and migration, except that cell division is arrested. The nucleus is located in a central cytoplasmic bridge as indicated previously. Where the bridge is in contact with the parietal cytoplasm, the cell wall becomes modified and it is here that it develops the characteristics of a preinfection thread if the thread from the infection pocket makes contact. After meristem formation, cells

(*Continued*)

**BOX 5.1 (Continued)**

steadily fail to show activation by the Nod Factors as the nodule grows and eventually emerges through the root surface.

The infection thread traverses several cells in the root cortex to reach the newly dividing cells below the nodule meristem. As infection threads enter this region, the bacterial cells are released into cells from wall-less branches of the infection threads into the plant cytoplasm and enveloped by a plant membrane, the peribacteroid membrane, derived from the host plasma membrane. The bacteria then enlarge and differentiate into nitrogen-fixing forms that are known as bacteroids. These bacteroids, with the surrounding membrane, are known as symbiosomes. It is in these structures that the symbiotic nitrogen fixation takes place. The mature nodule also incorporates two or more peripheral vascular bundles that converge toward the nodule apex and provide the means for exchange of nutrients between plant and nodules (Guan et al., 2013).

From these brief accounts of the formation of these two symbioses, it is evident that there is considerable similarity in the development of the symbiosis between the contrasting microbial symbionts – fungi and bacteria – and the host legume plant (Gianinazzi-Pearson and Dénarié, 1997). For example, although the signal compounds from the host plant are specific, strigolactones for AMF and flavonoids for rhizobia, the response signal from both microbial symbionts is a LCO and there appears to be some commonality in the nature of the receptors used by the host plant. Furthermore, both require the development of an infection thread-like structure for the symbionts to enter or pass through cells without penetrating the host plasma membrane (Kistner and Parniske, 2002). Some of the plant molecules associated with early events of rhizobia and legumes interactions have been located in AM symbiotic structures. For example, in pea, plant proteins and glycoproteins in the matrix surrounding bacteria in nodule infection threads are present in the host wall material around arbuscule hyphae. Oligosaccharides or glycoconjugates of the plant-derived membrane or interfacial matrix around the bacteroids in nodule cells are common to the peri-arbuscular membrane and arbuscule interface (Gianinazzi-Pearson et al., 1991b, 1996; Perotto et al., 1994). Both symbioses are inhibited by the phytohormone ethylene (Guinel and Geil, 2002) and there is evidence for the involvement of several phytohormones in the development and maintenance of the symbiotic structures, both pre- and postinfection (Hirsch et al., 1997; Downie, 2010). In addition, naturally occurring and chemically induced single gene mutants of pea (*Pisum sativum* L.) and faba bean (*Vicia faba* L.) are not able to form either functional root nodules with appropriate rhizobia or mycorrhizas with AM fungi (Duc et al., 1989). However, one major difference is the high level of specificity for the rhizobia partner shown by a host, whereas that is less obvious for AMF. Nevertheless, it is commonly considered that the development of the symbiosis with rhizobia involved the exploitation of the preexisting signaling system for mycorrhiza formation (Roberts et al., 2013). One consequence is the possibility that the two microbial symbionts could be competitive over the formation of a symbiosis, either for sites of infection or for plant resources, or, alternatively, that a tripartite symbiosis involving both microbes could be synergistic.

### *Interactions Between AMF and Rhizobia Affecting the Growth of the Legume Host and N Fixation*

Smith and Bowen (1979) concluded from their study on the effect of temperature on the colonization of *M. truncatula* that there was no competition for infection sites between native AM fungi and *S. meliloti*. From a metaanalysis of results from 20 papers published before 1983, Cluett and Boucher (1983) reported that the presence of mycorrhizal infection significantly increased nodulation in a range of legumes relative to those grown in the absence of AM fungi. However, the growth of the host legumes was greater when mycorrhiza were formed but differences were not significant if calculated as a function of plant dry weight. In one case (data from Bethlenfalvay et al., 1982) considered by Cluett and Boucher (1983), nodulation was significantly reduced by mycorrhiza formation. The legume was bean (*Phaseolus vulgaris* L. cv. Dwarf) and the two microbial symbionts were *Glomus fasciculatum* Gerd. and Trappe for the AM fungus and the rhizobia was *Rhizobium phaseoli*. The experiment considered the effect of the addition of P in the form of hydroxyapatite to the rooting medium, a mixture of perlite and sand, on the formation of nodules and mycorrhiza. Bethlenfalvay et al. (1982) concluded that when soil P greatly limits the overall growth of the tripartite symbiosis, there was competition between the two microbes through limitations in the supply of P. At levels of P that allow extensive hyphal development and in the absence of sources of N other than the symbiosomes, it is competition for carbohydrates between the two microbes that negatively impacts nodule development (Fig. 5.1). Importantly, at intermediate levels of P supply from the soil, the tripartite symbiosis is very effective for each participant.

> *The relative timing of colonization by AMF and rhizobia is important in the formation of the tripartite symbiosis but interaction between the symbionts does not take place within functioning root nodules.*

In another experiment this time using a 2:1 mixture of silt loam soil and sand, Bethlenfalvay et al. (1985) investigated the tripartite symbiosis between soybean (*G. max* [L.] Merr.), the AM fungus *Glomus mosseae* (Nicol. and Gerd.) Gerd. and Trappe, and the rhizobia *Bradyrhizobium japonicum*.

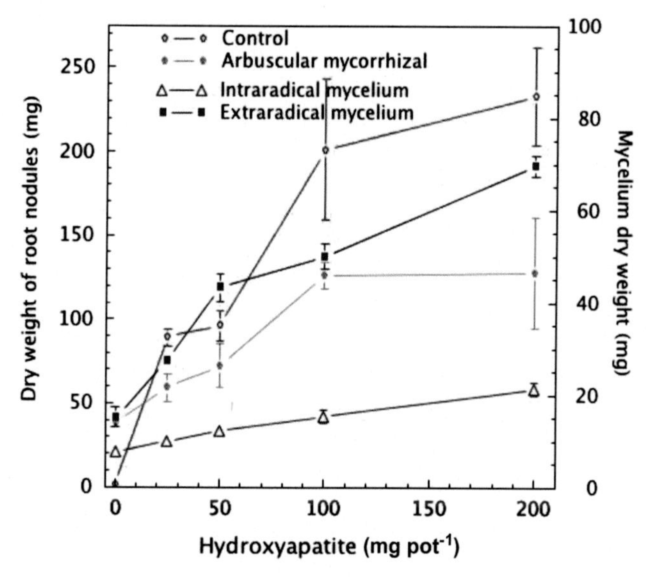


**FIGURE 5.1** Effects of adding phosphorus in the form of hydroxyapatite on the formation of nodules on bean roots by *R. phaseoli* in the absence (Control) and presence of arbuscular mycorrhiza. The effects of P on the growth of the inter- and extraradical mycelium is also shown. Without addition of P to the soil, the mycorrhiza supported a significant development of nodules, whereas there was almost no development in the Control treatment. With applications of 100 mg hydroxyapatite or more, there was no significant increase in nodule formation in either treatment but the mycorrhizal plants formed a little more than half of those present in Controls. The intraradical mycelium increased threefold with the addition of P, whereas the increase in extraradical mycelium (ERM) was almost fivefold. Root dry weight in the mycorrhizal plants increased by 26%, whereas in the Control treatment the increase was 45% (data not shown). *Results from Bethlenfalvay et al. (1982).*

The microbial symbionts were either applied singly (Treatments $F_1$ or $R_1$) or simultaneously (Treatment $F_1R_1$) to the roots of soybean plants at the start of the experiment. For those plants receiving only AM fungus, N as ammonium nitrate was added to the soil after 10 days. Plants receiving only rhizobia had P as potassium phosphate added to the soil, also after 10 days; others had no additional P (Treatment $R_1^0$). Other plants received both minerals N and P (after 10 days) but no symbiotic microbe (Treatment NS). After 20 days the soil was leached to remove minerals N and P before the "missing" microbe was applied to some of the plants that had previously received only one (Treatments $F_1R_{20}$ and $F_{20}R_1$), some plants that had not received a microbe now received both the AM fungus and the rhizobia (Treatment $F_{20}R_{20}$) or just rhizobia (Treatment $R_{20}$). Some plants continued as Treatments $F_1$, $R_1$, $R_1^0$, and others as Treatment NS. At harvest after 50 days growth, plants that received minerals N and P or either N or

P in conjunction with the complementary microbe (Treatments NS, $F_1$, and $R_1$), all had the same total dry weight and equal to that of plants that were inoculated with both microbes 20 days after the start of the experiment (Treatment $F_{20}R_{20}$). Similarly total dry weight of plants inoculated with one or both microbes at the start of the experiment (Treatments $F_1R_{20}$, $F_{20}R_1$, and $F_1R_1$) was similar but smaller than the previous group of treatments. Plants inoculated with *B. japonicum* at the start of the experiment but received no N or P (Treatment $R_1^0$) had the smallest dry weight. Nodule dry weight in Treatment $F_1R_1$ was greater than that in Treatment $F_1R_{20}$ but similar to that in treatments $F_{20}R_1$ and $F_{20}R_{20}$. Apparently nodulation was adversely affected if colonization by AMF occurred an extended period before that of rhizobia. Similarly if colonization by AMF took place well after nodulation (Treatment $F_{20}R_1$), then the dry weight of fungal material produced was significantly reduced compared with when the two inoculums were applied at the same time (Treatment $F_1R_1$). Delaying AMF colonization, even if it took place at the same time as the introduction of rhizobia (Treatment $F_{20}R_{20}$), reduced AMF colonization compared with early colonization (Treatment $F_1R_1$). These results were consistent with the previous study and strongly suggest that there is some level of competition between the two microbial symbionts. This is further supported by the fact that the greatest nodulation or AM fungus colonization occurred in the absence of the other symbiont and suggests that resource availability is important.

Smith et al. (1979) investigated the tripartite symbiosis in subterranean clover (*Trifolium subterraneum* L.) inoculated with *Rhizobium trifolii* and indigenous AMF. In soil with a similar supply of available nutrients, the growth of mycorrhizal plants was greater than nonmycorrhizal plants grown in autoclaved soil to which a soil filtrate was added. The purpose of the filtrate was to rebuild the bacterial population. In a 9:1 mixture of autoclaved and fresh soil, shoot weight of *T. subterraneum* was intermediate between that of mycorrhizal and nonmycorrhizal plants. In this case, mycorrhiza formation was much slower than in the fresh soil, which was attributed to a much smaller level of inoculum. Nodule development of mycorrhizal plants in fresh soil was greater, with more and larger nodules than on nonmycorrhizal plants. In the soil mixture nodules tended to be smaller than those in the fresh soil but after 8 weeks the number of nodules was greater and they were widespread on lateral roots as well as on the taproot of *T. subterraneum*. In contrast, in an experiment using soil with limited nutrient availability, Smith et al. (1979) observed that growth of nonmycorrhizal plants was better than that of mycorrhizal plants but the reverse was true for nodulation based on volume and activity of nodules (nonmycorrhizal plants formed slightly more but much smaller nodules). Here poorer growth in mycorrhizal plants most likely resulted from increased carbon demand to support the mycorrhiza and nodules. Other potentially limiting factors can include the supply of trace elements (Smith and Daft, 1977) or photosynthate (Bethlenfalvay et al., 1982).

All these early experiments investigating the interaction of mycorrhiza and rhizobia in the tripartite interaction with legumes indicated the importance of rapid mycorrhizal infection for enhancing nodulation by rhizobia. Smith et al. (1979) pointed out that a delay in colonization could occur if the mycorrhizal inoculum applied consisted of spores or infected root segments. We will consider this further later in this section. Importantly the research was consistent in that irrespective of the outcome in terms of shoot growth, mycorrhiza development resulted in enhanced P inflow to the host plant and the concentration of P in tissue tended to be least in the shoot, greatest in nodules and intermediate in colonized roots. The issue of competition between the symbionts for carbon from the host plant is also worthy of further consideration, not least because it contributes to the assessment of the role of the AMF component in the tripartite symbiosis. Another area that has received relatively little attention is whether the two symbionts interact directly within nodules. Early reports indicated that AMF hyphae are not found in nodules but that has been challenged (e.g., Scheublin et al., 2004, and references therein) but in a subsequent paper Scheublin and van der Heijden (2006) concluded that only nonfunctional nodules were colonized. The presence of spores in colonized nodules suggested that the AMF were making use of the resources in the nodule material rather than contributing to nodule functioning (Scheublin and van der Heijden, 2006).

*A key feature of the AMF symbiosis is the exchange of carbon, in the form of sugars from the host, for phosphorus from the fungus. The AMF component of the tripartite symbiosis results in greater photosynthesis by the host, which may result from a larger leaf area or greater photosynthetic efficiency.*

Direct effects of the two microbial symbionts on the supply of photosynthate in the tripartite symbiosis between legume, AMF, and rhizobia were investigated by Kucey and Paul (1981, 1982). They found that the rate of carbon fixation per unit leaf area by *Vicia faba* increased by 13.9% in the tripartite symbiosis than in the absence of the microbial symbionts. The AM fungus (*G. mosseae*) utilized 4% of the C fixed by the host and the rhizobia (*R. leguminosarum*) used 12% of the fixed carbon in the tripartite symbiosis (6% in the absence of the fungal symbiont). Harris et al. (1985) reported a similar value for the proportion of host-fixed carbon used by the rhizobial symbiont *B. japonicum*, symbiotic with soybean and *G. fasciculatum* (Thaxter sensu Gerd). Kucey and Paul (1981, 1982) found that although nodule biomass was 18.6% greater in the tripartite symbiosis than in the absence of the AM fungus, the rate of nitrogen fixation per unit nodule weight remained the same. Brown and Bethlenfalvay (1987, 1988) compared the plant carbon exchange rate of leaves from soybean involved in a tripartite symbiosis with *G. mosseae* and *B. japonicum*, in a simple symbiosis with two microbial symbionts separately, or with no microbial symbiont. In the absence of *G. mosseae* plants were provided with minerals P and N was supplied if the rhizobia was not present. In the tripartite symbiosis, the carbon exchange rate per unit area of leaf in the two studies increased by 19.4% and 30.6% relative to controls with no microbial symbionts. The increase in carbon fixation in *M. truncatula* symbiotic with the AM fungus *Rhizophagus irregularis* BEG141 was ascribed by Adolfsson et al. (2015) to increased branching and leaf canopy rather than carbon fixation per unit leaf area. The results from these various experiments suggest that the AMF and rhizobia symbionts act as additional carbon sinks, which result in increases in carbon fixation by the host plant. When AM fungi are investigated separately, whether in legumes or nonleguminous plants, the increase in the sink size ranges from 4% (Kucey and Paul, 1981, 1982) to 20% (Jakobsen and Rosendahl, 1990; Peng et al., 1993) of the total photosynthate produced by the host. The evidence for plants, including legumes, suggests that the enhanced carbon fixation by the host is the result of a combination of increased photosynthetic area and photosynthetic rate per unit area (e.g., Miller et al., 2002).

In a metaanalysis, Kaschuk et al. (2009) concluded that both symbionts provided an additional carbon sink, which were additive not synergistic, and this was important in enhancing photosynthesis in the host plant. Larimer et al. (2010) also reported additive effects of AM fungi and rhizobia following a metaanalysis of published material. In their investigation of the growth of the prairie legume *Amorpha canescens*, Larimer et al. (2014) reported that the AM fungi *G. mosseae* and *G. claroideum* increased the number and mass of nodules, even in soils where inorganic N adversely affected nodulation. However, the presence of rhizobia decreased colonization by AM fungi. Depending on the soil nutrient environment, the growth of the legume was enhanced by a particular combination of AM fungus and rhizobial strain. However, a contrasting combination could be more beneficial to the host plant in a different environment. For example, plants mycorrhizal with *G. mosseae* alone in combination with rhizobial strain 2 produced the best growth when P was added to the soil but when the mycorrhizal inoculum was a mixture of ~38% *G. mosseae* and 62% *G. claroideum*, the best growth for the combination with rhizobial 2 was in soil with no additional P. Overall both inoculation with AMF or rhizobia increased biomass production in *A. canescens* compared with controls provided with P or N as mineral nutrients, respectively. However, the effect of developing a tripartite symbiosis was synergistic and not simply additive. To explain the contrast between these results and the conclusions of Kaschuk et al. (2009) and Larimer et al. (2010), Larimer et al. (2014) suggested that many previous experiments had focused on annual plants important in agriculture, where insufficient time was available for the tripartite symbiosis to become synergistic.

*The tripartite interaction is greatly enhanced if AMF colonization is initiated from an intact mycorrhizal mycelium network. The ERM network can be a transport highway for nitrogen between a legume and a nonlegume host protection of the symbiosis against abiotic stress includes the defense of the rhizobial bacteroids.*

Goss and colleagues considered the importance of the speed of establishing the tripartite symbiosis to its efficacy (see Section 6.1 for a general account). Building on the work of Miller on the establishment of effective AM mycorrhiza in maize (Miller, 2000), detailed consideration was given to the potential of a preformed ERM as a primary inoculum (see Section 6.2.3) instead of spores or colonized root fragments, which can be slow to colonize (Smith et al., 1979). This was achieved, either by sieving the Canadian silt loam soil or leaving it undisturbed after growing a mycotrophic ERM developer plant. Instead of using sterilized soil and inoculating a laboratory strain or strains, Goss and coworkers followed Smith et al. (1979) in using the indigenous AMF population in the soil and generated two levels of inoculum potential: one level having spores, colonized root and an intact ERM, the other comprising spores and infected root fragments. They used a commercially available, peat-based inoculum of *B. japonicum* strain 532 C in their work with soybean. In a greenhouse experiment, Goss and de Varennes (2002) showed that the presence of the ERM in the inoculum resulted a faster colonization by both AMF and rhizobia compared with the presence of spores and root fragments. For example, 10 days after emergence arbuscules were present in 56% of root length and 14 nodules had been produced per plant when ERM was present compared to 14% of root length and 8 nodules when it was not. Importantly, AMF colonization increased in both treatments to podfill, but was always greater in plants with ERM in the inoculum. However, the number of nodules was unchanged after 23 days and similar in both treatments but the dry weight was consistently greater where ERM was in the inoculum and this was reflected in a threefold difference in $N_2$-fixation at podfill. Plant dry weight was similar in the two treatments at 10 and 23 days after emergence but by podfill (49 days after emergence) dry weight of plants having ERM present at sowing was 42% greater than where it was not. In contrast the number of trifoliolate leaves was greater 10 and 23 days after emergence, when ERM had been present, but inoculum potential made no difference to leaf number by podfill. The content of P in plants declined similarly in both treatments until 10 days after emergence, after which uptake started in the plants with ERM in the inoculum. However, uptake of P was delayed until 23 days after emergence in plants where there was no ERM in the inoculum. Total N acquisition by soybean at podfill was greater in plants with ERM in the inoculum but the concentration in the shoot was less than in plants where there had been no ERM. The faster AMF colonization from an intact ERM resulting in earlier nodulation of soybean was confirmed in a similar soil type but with a greater P content (Antunes et al., 2006b). However, in this field experiment, where rotary tillage to 10 cm was used to disrupt ERM or the soil was left undisturbed before soybean was planted with a no-till seeder, the effects did not result in any measurable differences in the soybean plants at podfill.

In a greenhouse experiment similar to that of Goss and de Varennes (2002) but using the annual medic, *M. truncatula*, instead of the grain legume, soybean, with *S. meliloti* as the rhizobia, de Varennes and Goss (2007) also found a more rapid colonization by indigenous AMF from a Portuguese clay soil when the AMF inoculum included ERM as well as spores and infected root fragments. However, by podfill, no differences in colonization remained. At 14 days and 29 days after emergence (flowering) there were no differences in shoot weight or nodule numbers between treatments but by podfill shoot weight and nodule size were greater where ERM was present in the AMF inoculum. Both the concentration and content of P in plants was greater throughout the experiment in the treatment with the ERM in the AMF inoculum. A greater proportion of N in the plants at podfill had been derived from the atmosphere where earlier AMF colonization had taken place.

Earlier, Kadir (1994) investigated the tripartite symbiosis in soybean under greenhouse conditions following different applications of P to the soil. The mycorrhiza were formed by indigenous AMF and either free-living rhizobia or the *B. japonicum* strain 532 C. The main treatment comparison was between an AMF inoculum that did or did not contain intact ERM. In soil containing intact ERM, plants grew faster and had a greater trifoliolate leaf area after 4 weeks than those infected by inoculum containing only colonized root fragments and spores. The difference in leaf area persisted to the time the plants were harvested at the end of podfill and was reflected in a difference in plant dry weight (Fig. 5.2A). By that time there were no significant differences between the main treatments in root colonization by AMF hyphae or arbuscules but the proportion of root length containing vesicles was significantly greater when an intact ERM was present in the inoculum (Fig. 5.2B). The difference in vesicle colonization was more consistent in the nodulating isoline than in the nonnodulating isoline (Fig. 5.2C). In contrast to the small effects on AMF colonization when ERM was the key propagule in the inoculum, the effects on rhizobia were much greater resulting in 38% more nodules than when only colonized root fragments and spores were present. The addition of phosphate to the soil produced a reduction of 5%–15% in the intensity of colonization by AMF (Fig. 5.2B) but increased the colonization by rhizobia, as assessed by the total weight of nodules (Fig. 5.2D). However, the negative effects on the proportion of root length containing arbuscules were large, but only at a concentration of 80 mg P $kg^{-1}$, when the

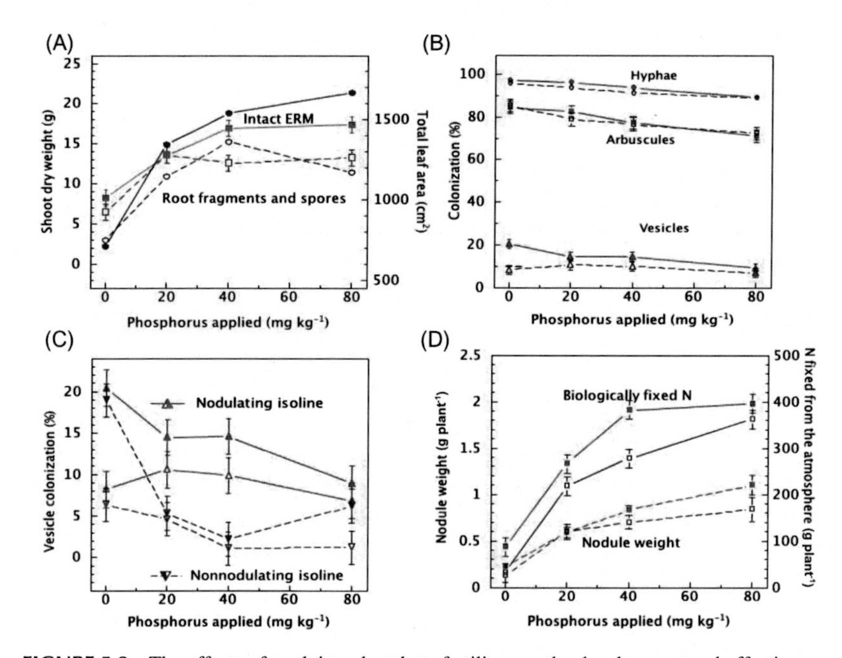


**FIGURE 5.2** The effects of applying phosphate fertilizer on the development and effectiveness of the tripartite symbiosis between soybean, indigenous AMF, and *Bradyrhizobium japonicum*. (A) Variation in shoot dry weight (red markers) and total leaf area (blue markers) at podfill in plants colonized using inoculum with extraradical mycelium (ERM) kept intact (solid markers) or made up mainly of root fragments, spores, and disrupted ERM (open markers). The application of phosphorus in excess of 20 mg $kg^{-1}$ to disturbed soil without an intact ERM was not beneficial to growth but plants colonized from intact ERM showed a significant response up to 40 mg $kg^{-1}$. (B) At podfill there was no effect of inoculum type on colonization, except for the concentration of vesicles within the roots, where the effect of the presence of an intact ERM was significant at $P < 0.001$. Red markers, ERM intact; blue markers, ERM disrupted. Negative effects on colonization of applying phosphorus were small. (C) The negative impact of phosphorus on colonization was greater in soybeans that were genetically incapable of establishing a viable symbiosis with Rhizobia and could not form nodules (blue markers) than in the nodulating isoline (red markers). The benefit to colonization from an inoculum containing intact ERM (closed markers) was also less consistent in the nonnodulating isoline. (D) Both colonization by Rhizobia, as indicated by nodule weight (dashed lines), and biological nitrogen fixation (solid lines) were enhanced by the presence of intact ERM (closed markers) when soybeans were planted compared with those colonized from spores, root fragments and disrupted ERM (open markers). Source*: Data from Kadir (1994).*

value was still in excess of 70%. By podfill the main treatments had no significant effect on the concentration of P or N in the shoots but the proportion of N in the plant resulting from biological fixation in the nodules was greater where AMF colonization took place in the presence of intact ERM and was also enhanced by the application of P (Fig. 5.2D). The relationship between nodule weight and N acquired by biological fixation was enhanced if the ERM was kept intact prior to planting the soybean and free-living wild-type rhizobia were available to establish functional nodules. Inoculating with the more effective 532 C strain further increased the N derived from biological fixation (Fig. 5.3).

The variation in the sensitivity of the tripartite symbiosis to added P in the soil, seen across the experiments discussed in this section, may also reflect the normal phosphate environment experienced by the AMF before the imposition of experimental treatments (Jasper et al., 1979).

One other important aspect of colonization by an existing ERM is that the potential exists for the new mycorrhizal plant to be linked to other plants. Enhanced transport between soybean and maize via a common ERM (van Kessel et al., 1985) indicated the potential for AMF to facilitate the transfer of N between legumes and grasses, and this was demonstrated by Haystead et al. (1988) with the transfer of N from white clover (*Trifolium repens* L.) to ryegrass (*Lolium perenne* L.). However, at least in laboratory microcosms, if more than one potential recipient host plant is linked to the same ERM network, there can be considerable competition between them (Walder et al., 2012).

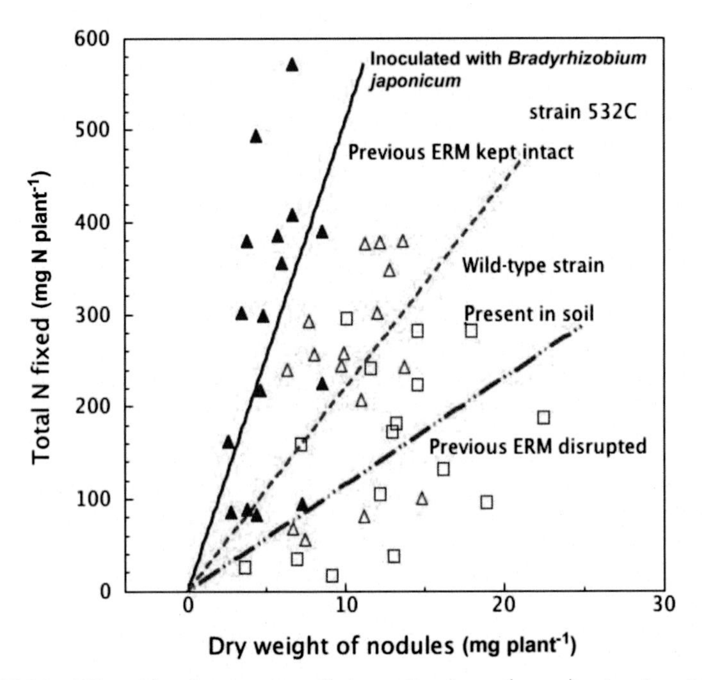


**FIGURE 5.3** Effect of keeping the extraradical mycelium intact (— —) rather than disrupted (—— · ·) prior to sowing soybeans on the effectiveness of nodules colonized by free-living wild type rhizobium and the added benefit from inoculation (———) with the more effective strain 532C. Source: *Data from Kadir (1994).*

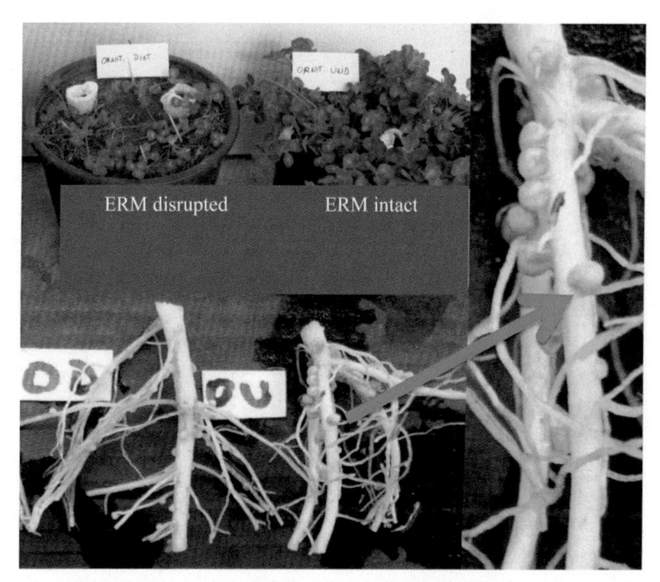


**FIGURE 5.4** Pots of *Trifolium subterraneum* L. 6 weeks after planting in soil containing 22.6 mg Mn $kg^{-1}$. Left, soil was sieved after the growth of the previous plants (*Ornithopus compressus* L.) and the roots were cut into sections and mixed back into the soil before subterraneum clover was sown. Roots were colonized by indigenous AMF from spores, colonized root fragments, and short pieces of disrupted extraradical mycelium (ERM). Note the sparse formation of nodules. Right, prolific growth of *T. subterraneum* in undisturbed soil, colonized by indigenous AMF from intact ERM (associated with *O. compressus*) and spores. More large nodules were formed on the main axis (arrow points to nodule on enlargement of main axis).

## *Resilience to Stress in the Tripartite Symbiosis*

It is well established that abiotic stress, such as from toxic ions, can directly affect $N_2$-fixation. The impact of toxic levels of Mn can greatly inhibit plant growth but a number of researchers (e.g., Dobereiner, 1966; Evans et al., 1987; de Varennes et al., 2001) found that this was greater, when legume plants were dependent on symbiotic $N_2$-fixation rather than on mineral N.

The effects of Mn can reduce the formation of root nodules in terms of numbers and size as well as the rate of symbiotic $N_2$-fixation (Evans et al., 1987; DeHaan et al., 2002). Alho et al. (2015) showed that, in *T. subterraneum* L., colonization by AMF from inoculum containing intact ERM was very effective in protecting plants from Mn (Figs. 5.4, 5.5). Shoot dry weight was up to 3.3 times larger after 21 days and, as indicated in Fig. 5.4, a maximum of 16.2 times greater after 42 days relative to colonization from spores or infected root fragments. The protection seemed to be associated with a smaller concentration of Mn in the plant roots (Fig. 5.6). As Mn in the roots increased, nodule dry weight decreased and so did the N content of shoots (Fig. 5.7). The decrease in Mn in the roots as a result of enhanced AMF

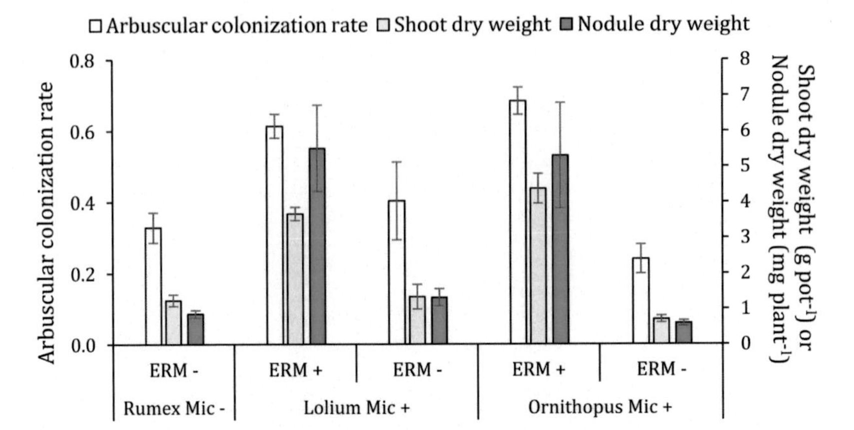


**FIGURE 5.5** Effects of developer ERM, presence and integrity, on colonization rate by indigenous AMF (based on arbuscule formation) and dry weight of shoots and root nodules of *Trifolium subterraneum* L. 21 days after sowing. Mycorrhizae were initially formed in association with roots of two common Mediterranean weed species (*Ornithopus compressus* L. or *Lolium rigidum* Gaudin), both being mycotrophic (Mic +). A third plant, *Rumex bucephalophorus* L. is considered not to form mycorrhiza (Mic-) and hence provided a control for soil disturbance and the contribution of disrupted ERM. ERM intact – ERM +, ERM disrupted – ERM – . Source: *Based on Alho (2015).*

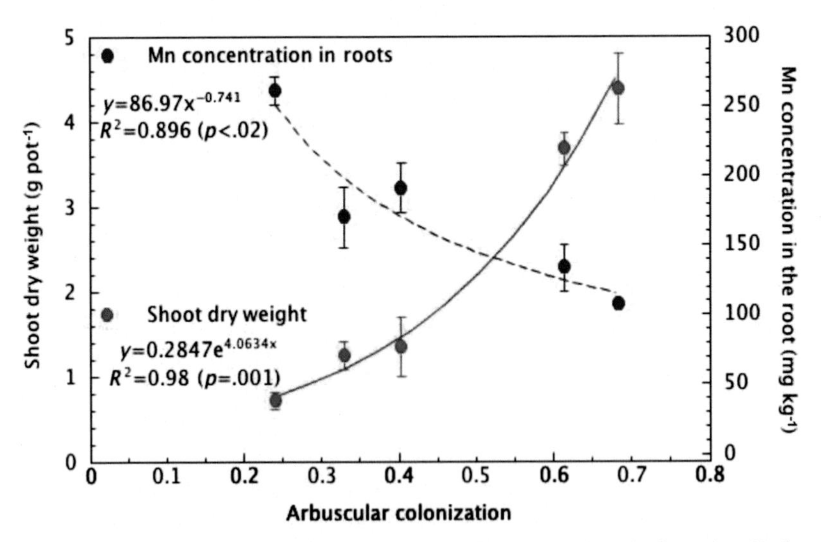


**FIGURE 5.6** Relationship between colonization rate, based on arbuscule formation 21 days after sowing, and shoot dry weight (- - -), and Mn concentration in the roots (——) of *Trifolium subterraneum* L. 42 days after sowing.

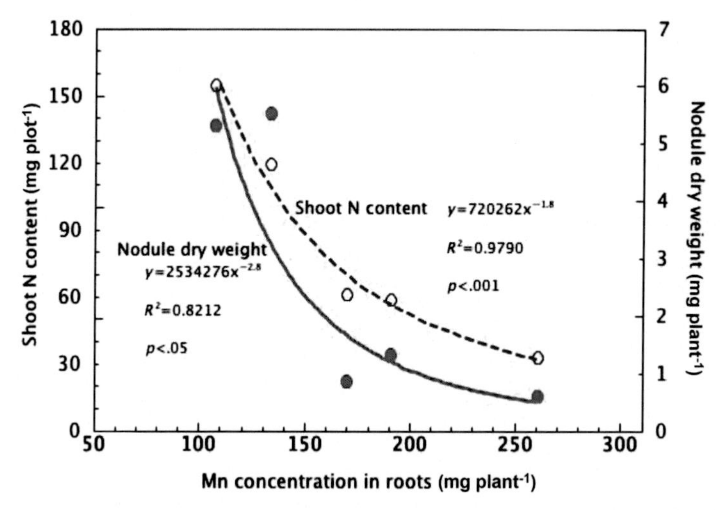


**FIGURE 5.7** Relationship between Mn concentration in the roots, shoot N content (— —) and Nodule dry weight (——) in *Trifolium subterraneum* L. 42 days after sowing.

colonization would explain the greater nodule dry weight, shoot N content, and dry matter production. Although at 21 days after planting, the proportion of root length containing arbuscules was up to 2.8 times that following colonization from colonized root fragments and spores, after 42 days the maximum difference between treatments was only 18% better when intact ERM was present. The protection granted by an enhanced AMF root colonization resulted in nodule weight ranging between 6.4 and 4 times greater when intact ERM was present in the soil than when it was not. As nodule weight is a good indicator of the effectiveness of the symbiosis between the legume and rhizobia, this suggests that the benefit of AMF to the tripartite symbiosis was through an effect on the microbial symbiont as well as on the higher plant (Fig. 5.7).

*Bacteria, both free-living and endophytic within the hyphae, can enhance the effectiveness of AM mycorrhiza, both by aiding colonization and increasing the effectiveness of protection of the plant host against biotic and abiotic stresses.*

## 5.1.2 Other Interactions With Bacteria

The ERM is commonly associated with bacteria that are attached to the surface of the hyphae (Scheublin et al., 2010) or even living as endocellular organisms within the cytoplasm of hyphae (Bonfante and Anca, 2009). With modern molecular techniques, the different taxonomic groups of bacteria found within the mycorrhizosphere have begun to be identified (Fig. 5.8).

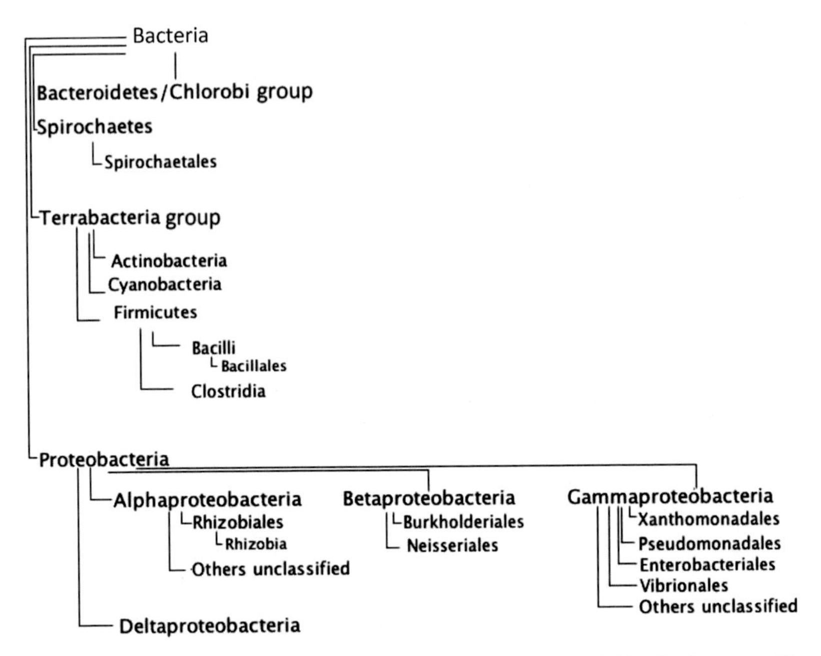


**FIGURE 5.8** Main bacterial groups considered to participate in activities in the mycorrhizosphere (based on Bonfante and Anca, 2009). Although alphaproteobacteria include the rhizobia that form the tripartite symbiosis with legumes, some members of the betaproteobacteria can also fix nitrogen from the atmosphere and have been identified as endobiotrophs (Cage, 2004; Leveau and Preston, 2008).

Toljander et al. (2006) observed differences in the mode of attachment to living and dead hyphae between taxa of bacteria, consistent with some being saprophytic and others, whose functioning depended on being more intimately involved with the living fungus. Some of the bacterial species that are found in close association with AM fungi have been shown to enhance the formation of mycorrhiza on receptive hosts (Garbaye, 1994). The specificity shown by individual bacterial strains for increasing AMF colonization in specific soils has resulted in the concept of *mycorrhiza helper bacteria* – MHB – (Garbaye, 1994). Some bacteria that apparently enhance the benefit to the host plant from the formation of a mycorrhiza are also known to be directly beneficial to host plants, being identified as plant growth-promoting rhizobacteria (PGPR). Possible mechanisms by which mycorrhiza formation can be improved have been distilled into four hypotheses:

1. MHB improve soil properties that are conducive to improved fungus colonization. Such properties include the creation of a more appropriate soil pH and the complexing of ions, especially those that can be toxic to growth, by siderophores.

2. MHB promote the germination of spores together with the growth and survival of mycelium. Both gaseous and small molecular weight compounds as well as sugars have been shown to affect germination or hyphal growth and branching (Artursson et al., 2006). MHB may also act through antagonism to or competition with other bacteria or fungi that are inhibitory of AMF activity and hence links to the first hypothesis (Artursson et al., 2006).
3. The presence of MHB improves the receptiveness of roots to initial interaction with the mycorrhizal fungi. Root branching stimulated by MHB will increase the intensity of root development and the likelihood of interception with mycorrhizal fungi. The formation of compounds, such as indole-3-acetic acid (IAA), by bacteria in the rhizosphere could not only modify root branching (though this still has to be established in MHB) but reinforce the impacts of the local inhibition of phytohormone transport during the initial responses to the recognition in the plant of the "myc factors" released by the fungus. One example is provided by the isolate UW4 of *Pseudomonas putida*. This MHB promotes mycorrhiza formation (increased root colonization and arbuscule formation) in cucumber by the AMF *Gigaspora rosea* by producing ACC deaminase that reduced the formation of the phytohormone ethylene, which is known to inhibit the colonization of roots and formation of fully developed arbuscules (Herrera-Medina et al., 2007). A mutant of *P. putida* UW4, lacking the deaminase, had no beneficial effect on mycorrhizal development (Gamalero et al., 2008).
4. MHB enhance the early stages of signal recognition between the host plant and the mycorrhizal fungus (Garbaye, 1994; Frey-Klett et al., 2007). Sanchez et al. (2004) compared the colonization of *M. truncatula* roots by the AMF *G. mosseae*, *S. meliloti*, and the MHB *Pseudomonas fluorescens* strain C7R12. *Glomus mosseae* increased the activity of 12 genes of *M. truncatula*, whereas *S. meliloti* upregulated only three of the same set of genes and down regulated five of them. *P. fluorescens* strain C7R12, which colonized the surface of root tips and grew between cells within the root cap and also entered some cells of the root cortex, increased the activity of seven of the gene set, consistent with the MHB causing a number of the same host responses as the AMF during colonization.

In contrast to MHB, some bacteria as well as a number of fungi can be antagonistic to AMF and are capable of parasitizing them (de Boer et al., 2005; Lee and Koske, 1994). Other bacteria use the exudates from AMF hyphae and this has been proposed as one mechanism that can affect the relationship of the fungus with specific bacteria associations (Andrade et al., 1997). The range of *bacterial mycophagy,* the active feeding of fungal material, covers extracellular necrotrophy, extracellular biotrophy, and endocellular biotrophy (Leveau and Preston, 2008). Necrotrophic actions involve the secretion by the mycophagous bacteria of proteins or toxins with a relatively small molecular weight, which increase the permeability of fungal hyphae or lyse them and inhibit fungal metabolism, leading to hyphal death and the release of metabolites that are used in bacterial growth. Extracellular biotrophy does not kill fungal hyphae, but the bacteria live in close proximity and may colonize the hyphal surfaces. The colonization can involve exopolysaccharides, surfactants, and fimbriae, reminiscent of the attachment of rhizobia. The bacteria may be able to modify the metabolism of the host, thereby increasing the exudation of nutrients from living hyphae or other fungal cells. Such biotrophs can tolerate or actively suppress production of antibacterial metabolites by the fungus. Endocellular biotrophy involves the absorption of nutrients from the fungal cytoplasm by bacteria located within living fungal cells, where they grow and multiply. AMF are likely to be subject to all three forms of bacterial mycophagy (Bonfante and Anca, 2009).

Endocellular bacterial biotrophs in AMF were first identified in microscopy studies, where they were reported as bacteria-like objects (BLO) (MacDonald and Chandler, 1981). Combining morphological and molecular techniques have allowed BLO to be identified (e.g., Bianciotto et al., 1996). The evidence indicates that the bacteria can be passed from one generation of AMF to the next as part of vegetative spore formation (Bianciotto et al., 2004). Much of our understanding of endotrophic bacteria in AMF has resulted from the study of Isolate BEG34 of *Gigaspora margarita* and its endobacterium "*Candidatus Glomeribacter Gigasporarum*," which has been used as a model system (Bonfante and Anca, 2009). The inability of the bacterium to be grown in culture outside of its host results in its designation of *Candidatus*. The endobacterium, which is a rod-shaped, gram-negative organism $\sim 0.8-1.2\,\mu m$ in diameter $\times 1.5-2.0\,\mu m$ in length, is found singly or grouped and often in protein-filled vacuoles within cells of the AMF, both in spores and hyphae. However, this organism is confined to the Gigasporaceae but a coccoid endobacterium (MacDonald et al., 1982) is more widely distributed across different groups of AMF and, although having a gram-positive cell wall, its ribosomal DNA indicates that it is related to Mollicutes (Naumann et al., 2010), which are common endophytic pathogens but do not produce a cell wall (Dybvig and Voelker, 1996). The two bacteria are able to coexist in the same fungal cell, although there is evidence that *Ca. G. gigasporarum* can be surrounded by a membrane of fungal origin, whereas the Mollicutes-related coccoid is free within the fungal cytoplasm (Desirò et al., 2014).

The transcription of the marker gene for cell division in *Ca. G. gigasporarum* is most active during the symbiotic phase of the AMF, particularly in the ERM (Bonfante and Anca, 2009). Treatment of *Gi. margarita* spores with strigolactone also stimulated the division of the bacterium. When the bacterium was selectively lost from the AMF hyphae, the elongation and branching of hyphae, following the application of root exudate, was greatly impaired, suggesting that the endobacterium had an important role in the preparation of the fungus for the formation of a symbiosis with a host plant (Bonfante and Anca, 2009). That would be consistent with the endobacterium being a MHB.

*AMF appear to interact with other fungi as well as soil fauna, with positive and negative impacts on the effectiveness of the mycorrhiza as well as some evidence of predation on fine hyphae.*

## 5.2 INTERACTIONS BETWEEN AMF AND OTHER FUNGI

There has been considerable testing of the hypothesis that AMF provide some protection to host plants against fungal pathogens. Fitter and Garbaye (1994) reported French research, which involved both ecto- and endomycorhiza, that found significant reductions in disease were achieved in 76% of cases studied. In many natural environments, particularly in sandy soils, AMF hyphae within roots have been associated with dark colored septate hyphae of fungi that are known to be or may be affiliated to the ascomycetes (Mandyam and Jumpponen, 2005). Despite many reports on the fungi there appears to be no indication of any interaction with AMF in the same plant. Yeasts, which produce vitamin $B_{12}$, have also been shown to act as mycorrhizal helpers, increasing root colonization and spore production, and also increasing the beneficial effects of the AMF on the host plant (Boby et al., 2008).

## 5.3 INTERACTIONS BETWEEN AMF AND SOIL FAUNA

Much of the interest has been focused on whether grazing by arthropods, such as mites and Collembola, have a serious impact on AMF, including reducing the size of the spore bank in the soil. The other area that has received considerable attention is the interaction with burrowing organisms, such as earthworms.

### 5.3.1 Interactions With Arthropods

In the herb *Geranium robertianum* L., growth was enhanced by the formation of a mycorrhiza but the introduction of the collembolan *Folsomia candida* into the soil caused a reduction in growth that was not directly related to grazing pressure (Harris and Boerner, 1990). McGonigle and Fitter (1988) reported broadly similar results. Such results suggested that grazing of the fine hyphae of the ERM was detrimental to the effectiveness of arbuscular mycorrhiza (Fitter and Garbaye, 1994; Hodge, 2000). Klironomos and Ursic (1998) found that in their study with *F. candida* the impact on the efficiency of the symbiosis was a function of grazing pressure. However, Larsen and Jakobsen

(1996) concluded that because the AMF hyphae were not the preferred diet of the Collembola, the real interaction between the two organisms was quite limited. Consequently in the specially developed experimental microcosm, which allowed *F. candida* to graze the extraradical hyphae of the AMF *Glomus caledonium* (Nicol. & Gerd.) Trappe and Gerdemann without roots being present, there was no effect of grazing on the growth of *T. subterraneum* over 6 weeks. However, thereafter the growth of mycorrhizal plants was slower than nonmycorrhizal plants. Adding yeast cells as an alternative food source for the Collembola increased hyphal length whether Collembola were present or not but a small apparent effect of grazing at 4 weeks was not detected thereafter (Larsen and Jakobsen, 1996). Gange (2000) concluded that on the balance of evidence available, there was no consistent indication that Collembola adversely impacted the beneficial effects of AMF on plants. In a microcosm experiment using maize (*Zea mays* L.) as host plant, Ngosonga et al. (2014) investigated the grazing of the collembolan *Protaphorura fimata* on the AMF *G. mosseae*. The growth of maize shoots was greatest where *P. fimata* was present in the compartment containing mycorrhizal roots or where it was also present in the compartment containing only AMF hyphae. The P content of roots was greatest where the Collembola were present in the root compartment of mycorrhizal roots but there was no effect of *P. fimata* on the acquisition of N. Grazing of mycorrhizal roots increased the dry matter invested by the AMF in the hyphal compartment, particularly if grazing was restricted to that area. Rather than developing spores or storage structures, the grazing of mycorrhizal roots encouraged greater hyphal exploration of the root compartment. The evidence from fatty acid profiles was that Collembola preferentially fed on soil bacteria but consumed more fungal material in compartments containing mycorrhizal roots. Overall the Collembola were beneficial to the mycorrhiza.

In a field microcosm study, Bakonyi et al. (2002) found that Collembola reduced the number of AMF spores in the soil under maize and *Festuca rubra*. At grazing densities between 0.2 and 0.4 adult Collembola $g^{-1}$ soil, the number of spores declined rapidly but at greater densities the decline was much less. At small grazing densities the colonization of roots by AMF increased, which was assumed to result from spores being transported by the Collembola to root surfaces. However, at grazing densities slightly greater than those that reduced spore numbers, there was a decline in the percentage of root length colonized by AMF.

### 5.3.2 Interactions With Earthworms

The presence of the earthworm *Lumbricus rubellus* in the soil increased the dry matter production of mycorrhizal plants of *Plantago lanceolata* but the presence of the animals did not affect the level of root colonization by indigenous AMF (Gormsen et al., 2004). The content of ERM in the soil, based on fatty acid-specific analysis for fungi was greatly enhanced by the presence of earthworms. Eisenhauer et al. (2009) also found that earthworms did not affect root colonization by AMF (*Glomus intraradices*) but also found no interaction between AMF and earthworms on the growth of representative grass, herb, and legume hosts. In contrast, Li et al. (2013) inoculated maize with AMF, which significantly improved shoot growth and maize yield; the addition of earthworms resulted in a further yield enhancement. There was an increase in the activity of alkaline phosphomonoesterase, in the soil and C and N in microbial biomass were also enhanced by the presence of both AMF and earthworms, whereas AMF reduced the availability of P.

## 5.4 CONCLUSIONS

Far from being a large diversity of competing organisms inhabiting the rhizosphere of plants, there is considerable evidence of both cooperation and synergism between groups concentrated around mycorrhiza. It seems that where the interaction between microbes and plants is of particular interest to the development of a sustainable agriculture, the relationship is carefully choreographed through complex signaling systems. Much of our detailed knowledge of the interaction between AMF, bacteria, and plants comes from legumes involved in a tripartite interaction that may also involve additional endophytic partners, which are only now being identified and their possible roles elucidated. The benefits of mycorrhiza in the tripartite interaction are best achieved when the host plant is colonized early, especially from an inoculum based on intact ERM. The outcome ultimately depends on the subsequent growing conditions. Very large improvements in growth can result, at least in the presence of abiotic stresses. The interactions between AMF and other organisms is less well understood but most can have some benefit to the development of the mycorrhizal host plant.

Chapter 6

# The Significance of an Intact Extraradical Mycelium and Early Root Colonization in Managing Arbuscular Mycorrhizal Fungi

## Chapter Outline



Much of the early interest in the extraradical mycelium (ERM) of arbuscular mycorrhiza (AM) focused primarily on its function in the acquisition of nutrients, mainly P, but it can also be very important in the provision of several other services for the symbiosis and for the ecosystem. These services include the colonization of new root systems, foraging for other nutrients, provision of a communications network for the host plants and the development and maintenance of soil structure. The ERM can therefore provide opportunities that can and should be considered within ecosystem management. In this chapter we address the most relevant characteristics and potential roles of the ERM and consider how management techniques can modify the development and functioning of the AM within the soil environment.


**Functional Diversity of Mycorrhiza and Sustainable Agriculture.**
**DOI: http://dx.doi.org/10.1016/B978-0-12-804244-1.00006-X**

## 6.1 IMPORTANCE OF EARLY ARBUSCULAR MYCORRHIZAL FUNGI COLONIZATION

AM are the most wide spread mutualistic symbioses between plants and soil fungi – specifically arbuscular mycorrhizal fungi (AMF). The symbiotic fungal partner is an obligate biotroph, deriving all its carbon from the photosynthetic activity of the host plant, which in return can benefit from a wide range of advantageous provisioning of resources and protection from stress by the fungus. Resource provisioning comes directly from the acquisition of soil nutrients, particularly those less mobile in soil, and beneficial interactions with other organisms and indirectly from the improvement of soil conditions (see Chapters 3 and 5). Stress protection involves alleviation of the impact of both biotic and abiotic factors.

It is axiomatic that early root colonization is crucial if the potential benefits from AM are to be optimized in the host plant. This is not only because of the obvious opportunities resulting from a better nutrient supply if the benefits start at the beginning of the vegetation cycle – the same being true for the ability to stand up against possible biotic or abiotic stresses – but also as it offers the possibility of precocious activation of plant defense mechanisms, both locally and systemically (Cameron et al., 2013; Khaosaad et al., 2007). Soilborne fungal pathogens and plant-parasitic nematodes occupy similar root tissues as AM. These organisms can compete for space in the root if colonization takes place at the same time. Therefore the number of AMF infection sites on a particular root system can determine the extent of pathogen ingress (Harrier and Watson, 2004).

> *It is the earliness of AMF colonization, and not necessarily its extent, that is fundamental to optimizing benefits for the host plant.*

It is not simply the level of colonization (the extent of internal colonization as indicated by the presence or intensity of hyphae or arbuscules) itself (Jun-Li et al., 2010) but how fast it gets established (infectivity) that is the important trait to permit plants to take advantages of the AM symbiosis (Garg and Chandel, 2010). This could be anticipated, given that multiple processes, such as P acquisition and root branching, are initiated under plant genetic control following AMF colonization, starting with only a single or a very few infection sites (Harisson, 2005; Poulsen et al., 2005; Schweiger and Müller, 2015), and the systemic nature of many AM targeted mechanisms. Furthermore, differences among AMF species are likely to modify the relationship between colonization and provision of mycorrhizal benefits, owing to recognized functional diversity (Abbott and Robson, 1981; Abbott et al., 1995; Munkvold et al., 2004). Even so the AM effectiveness (increasing plant growth) and colonization rate are often positively correlated (Abbott et al., 1995; Brito et al., 2014; Alho et al., 2015).

Following the inoculation of barley with two different AMF species at 4 days intervals, Vierheilig (2004) observed that high levels of AMF root colonization resulted in strong suppression of further root invasion, and ascertained an analogous effect in the fact that only a well-established symbiosis could protect plants against soilborne pathogens. Early AMF colonization would also be particularly relevant if the stress to be faced by the plant was already present in the soil when the crop was sown and could be detrimental to growth immediately the seeds germinated, e.g., in the case of metal toxicity problems.

Several examples illustrate the reality of the impact of an early AM colonization and its rate. In 1988 Read and Birch reported that the first impact of AMF colonization was a significant increase in the P:N ratio over the 5 day period from the time the infection was fully established (20 days after planting), whereas up to that time the P:N ratio had declined because of the dilution of shoot P levels as the utilization of P in seed reserves outstripped P inflow. Goss and de Varennes (2002) showed that colonization rate directly affected shoot P content in soybean. Another example was reported by Abbott and Robson (1981); in their study the fresh weight of shoots of subterranean clover growing in P-deficient soil was very closely related to the percentage of root length colonized at an early stage of plant growth. Also a more rapid colonization of soybean by AMF resulted in earlier establishment of the symbiosis between the plants and *Bradyrhizobium japonicum*, which gave rise to greater N fixation (Goss and de Varennes, 2002; see Section 5.1.1). In a recent study, the prior inoculation of tomato seedlings at the nursery stage, using a mixture of four naturally occurring AMF species, resulted in a faster establishment rate and greater vigor, which was reflected in the prolonging of the harvesting period and resulted in a 25% yield increase in open field tomato cultivation (Vuksani et al., 2015). A similar preinoculation at the nursery stage of the North American grapevine, *Vitis rupestris* Scheele, with *Glomus intraradices*, helped prevent "black foot" disease, both in the nursery and in the vineyard (Petit and Gluber, 2006). Under adverse conditions of Mn toxicity in the soil, Brito et al. (2014) reported a strong positive relationship between dry matter production (2.5 times more) of wheat plants and the timing of AM colonization when comparing plants that were early or late colonized.

*AMF propagules – spores, colonized root fragments, and ERM – have different characteristics and are not equally infective. Spores are long-living resting structures and take longer to initiate new colonizations. Colonized root fragments harbor vesicles that are able to provide energy for hyphal re-growth. ERM, when intact, is commonly the most infective propagule.*

## 6.2 ARBUSCULAR MYCORRHIZAL FUNGI INOCULUM SOURCES

Spores, fragments of AM colonized roots and ERM are the possible inoculum sources, collectively termed propagules (Smith and Read, 2008). They are all able to start new mycorrhizal colonizations of plant roots, although the different propagule forms exhibit different colonization capabilities (Requena et al., 1996; Klironomos and Hart, 2002). The process of colonization comprises several steps, including spore germination, hyphal differentiation, hyphopodium formation, root penetration, intercellular growth, intracellular arbuscule formation (Plate 6.1 and Section 5.1.1) and nutrient exchange, and is largely conserved among different combinations of fungal and plant species (Garg and Chandel, 2010).

In the presence of a mycotrophic plant, a complex biochemical dialogue toward recognition begins to takes place between the root system and the fungus (see Section 5.1). This dialogue is based on the sequenced emission and recognition of biochemical signals by both symbionts and prevents the full expression of plant defense mechanisms, allowing the symbiosis to become established (Harrison, 2005). Initially it was thought that phenolic compounds, in particular flavonoids, were especially important in the communication process, by analogy to the rhizobium–legume interactions. However, a major step forward in deciphering the molecular cross-talk in the AM symbiosis was the identification of a strigolactone as the host plant initiator of the dialogue present in root exudates that induces hyphal branching (Akiyama et al., 2005; Xie et al, 2010). Strigolactones also stimulate the genes that encode Myc factors and their release into the soil. The Myc factors are chitin oligomers and lipochitooligosaccarides that can interact with receptors in the epidermal cells of the host roots, thereby starting processes in the plant to allow the establishment of the symbiosis (see Box 5.1). Nonmycotrophic plants fail to establish this dialogue because they do not produce the right signaling compounds, such as hyphal branching factors (Buee et al., 2000), or do not produce them in the necessary amount to allow the AMF to undergo the colonization process (Harrison and Dixon, 1993) or even resist invasion (Bonfante-Fasolo and Perotto, 1992), clearly demonstrating that the host plant plays a key role in orchestrating the AM infection process.

### 6.2.1 Spores

Spores are resting structures produced by hyphae in the soil and the only plant-independent phase of the AMF. For a long time their morphology was the main trait to differentiate AMF species. With adequate temperature and water conditions, spores germinate producing a germination tube. This coenocytic hypha grows with marked apical dominance and can reach some

(A)

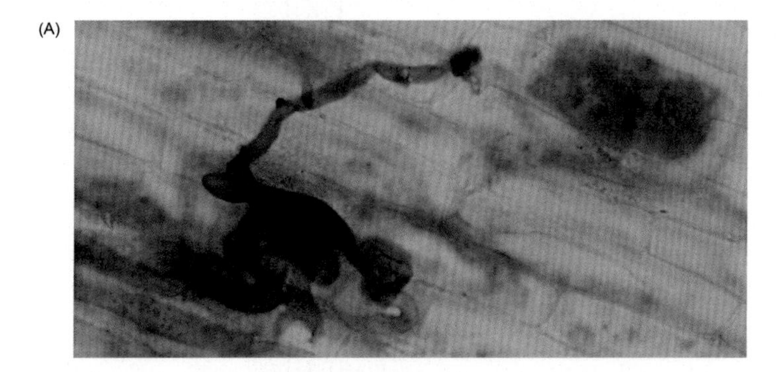

(B)

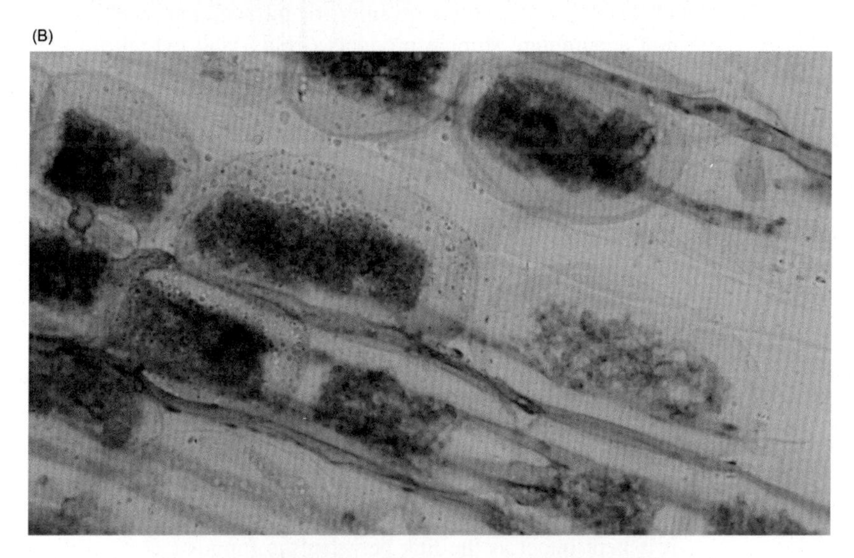

(C)

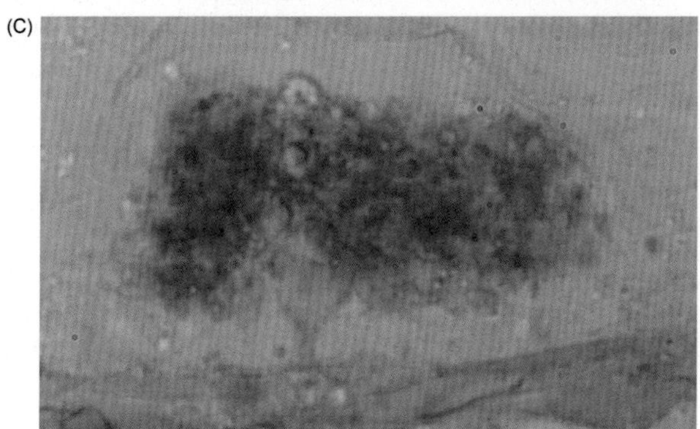

**PLATE 6.1** (A) Hyphopodium (200×); (B) hyphal intercellular growth and arbuscules (200×); (C) arbuscule (400×).

centimeters in length. If a compatible plant root is not present, growth ceases after about 2–4 weeks and the hyphae septate apically and the protoplasm retracts toward the spore (Mosse 1988; Logi et al., 1998), allowing its survival until an eventual restart. AMF spores can remain infective for several years.

### 6.2.2 Colonized Root Fragments

Colonized living and dead root fragments can also start new AM colonization. Root tissues protect the fungus from environmental hazards until the time when new hyphae can grow out from the roots and colonize other plants (Requena et al., 1996). This is possible, largely because they enclose structures called vesicles (Plate 6.2), which are inter- or intracellular swellings in the root cortex and accumulate storage products (lipids) and cytoplasm that serve as an energy source.

Vesicles are initiated soon after the first arbuscules but continue to develop when the arbuscules senesce. Vesicles can develop thick walls in older roots and may function as propagules (Biermann and Linderman, 1983; Tommerup, 1984). The production and structure of vesicles varies between different genera of AMF. Root pieces colonized by *Gigaspora margarita* or *Gigaspora gigantea*, which do not generally form vesicles, are not able to set off new AM colonizations (Biermann and Linderman, 1983). Plants can be colonized more rapidly when inoculated with AM colonized root fragments than when inoculated with spores (Abbott and Robson, 1981).

### 6.2.3 Extraradical Mycelium

The extensive network of ERM developed in the soil by mycotrophic plants can be an important inoculum source for new AM colonizations. This is particularly true if it is kept intact as the link between the fungus and its carbon provider, the host plant, which is important for its functionality. The ERM is

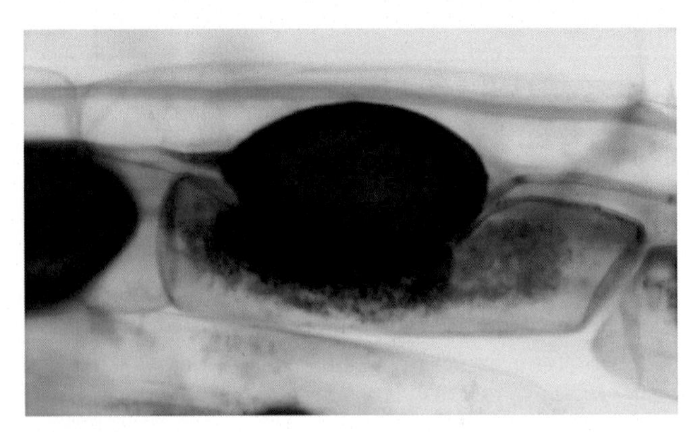

**PLATE 6.2** Vesicle (400×).

mainly composed of running hyphae and absorptive hyphae and the specialized hyphal architectures of AMF appear to be linked to the unique function of each hyphal type. Runner hyphae are capable of infecting new root segments, but absorptive hyphae have never been observed to act as units of infection and are classified as structures primarily involved in the acquisition of soil resources (Friese and Allen, 1991). However, this generalized architectural perspective of ERM was challenged by Dodd et al. (2000) who consider it more representative of the ERM of *Glomus* spp.

In undisturbed soil with supporting native vegetation, the hyphal network is more important than spores or colonized root fragments (McGee et al., 1997; Kabir, 2005) as inoculum provider. The likelihood that a plant root intersects an infection unit resulting from an ERM network will be greater and besides it takes longer for spores to germinate and make contact with roots as opposed to runner hyphae infecting from a well-developed ERM (Jasper et al., 1989a; Klironomos and Hart, 2002). The greater AM colonization initiated by an intact ERM is well documented both in pot experiments (Jasper et al., 1989a; Fairchild and Miller, 1990; Miller et al., 1995a; Martins and Read, 1997) and in vitro studies (Voets et al., 2009) as well as in natural (Read et al., 1976; Read and Birch, 1988), and agricultural ecosystems (Kabir et al., 1997a; McGonigle and Miller, 1996; Kabir et al., 1998; Galvez et al., 2001; Castillo et al., 2006a). In addition, earlier and faster colonization has also been identified in the presence of intact ERM (Brito et al., 2012a, 2013b; Goss and de Varennes, 2002).

There is some evidence that the biochemical recognition dialogue between AMF and the host plant leading to a functional mycorrhiza does not follow the same pattern, depending on whether the inoculum source is based on spores or intact ERM. David-Schwarts et al. (2001) isolated a tomato mutant, M126, that is able to resist to colonization in the presence of *G. intraradices* spores. However, normal colonization of M126 was evident when mutant plants were grown together with mycorrhizal-inoculated wild-type plants in the same growth medium, revealing a genetic controlled step in the AM colonization process, which significantly affected key aspects of premycorrhizal infection stages. Also Brito (unpublished data, 2014) observed that even very poorly mycotrophic plants can be colonized if ERM is the main inoculum source. In a pot experiment, *Rumex bucephalophorus*, a scarcely mycotrophic plant, when grown after the highly mycotrophic plant (*Ornithopus compressus*) and in the presence of an intact ERM, was four times more colonized and had a significantly greater shoot development, than when it was grown after a nonmycotrophic plant (*Silene galica*) and therefore the main source of propagule were spores (Figs. 6.1 and 6.2, treatments OR and SR, respectively). When grown following *S. galica*, *R. bucephalophorus* was only slightly colonized (2%). Similar results were observed by Püschel et al. (2007) who argued that because ERM is such a powerful

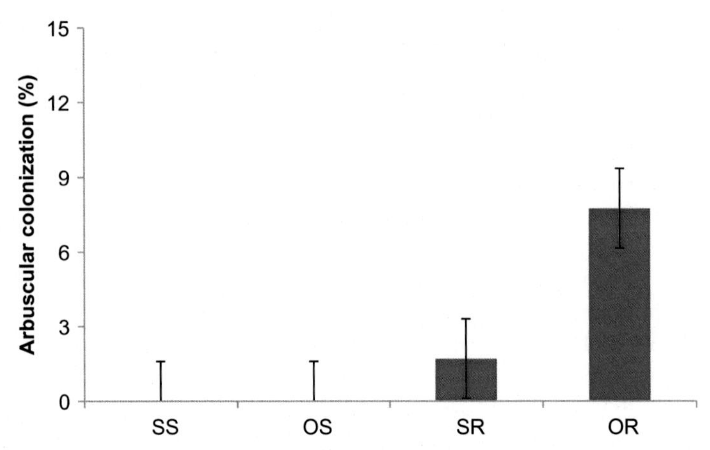


**FIGURE 6.1** Arbuscular colonization (%) of the second plant in a succession, grown in undisturbed soil. ERM was the main source of AMF propagule when Ornithopus was the first plant in the succession. S: *Silene galica* (nonmycotrophic plant); R: *Rumex bucephalophorus* (scarcely mycotrophic pant) and O: *Ornithopus compressus* (mycotrophic plant). The first letter of the labels corresponds to the first plant and second letter to the second plant in the succession (e.g., OR, is *Rumex bucephalophorus* preceded by *Ornithopus compressus*).

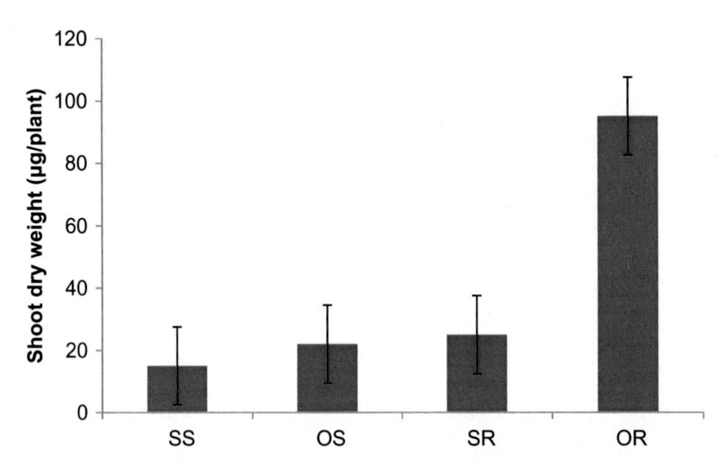


**FIGURE 6.2** Shoot dry weight (μg $plant^{-1}$) of the second plant in a succession, grown in undisturbed soil. ERM was the main source of AMF propagule when Ornithopus was the first plant in the succession. S, *Silene galica* (nonmycotrophic plant); R, *Rumex bucephalophorus* (scarcely mycotrophic plant) and O, *Ornithopus compressus* (mycotrophic plant). The first letter of the label corresponds to the first plant and second letter to the second plant in the succession (e.g., OR, is *Rumex bucephalophorus* preceded by *Ornithopus compressus*).

form of inoculum that even plant species usually not hosting mycorrhizal fungi could be colonized.

Whether a few separate types of signals are involved or morphological response are simply dose-dependent still needs to be elucidated and whether the existence of signals at the preinfection stages is more important for AM fungal infection by spores than by vegetative hyphae has likewise, yet to be determined.

> *The great potential of ERM as an inoculum source can be available and even enhanced in agronomic systems. Environmental conditions and AMF diversity together with the agronomic history of a particular site are key factors to take in consideration.*

## 6.3 ERM AS AN EFFECTIVE ARBUSCULAR MYCORRHIZA INOCULUM SOURCE FOR FIELD CROPS

The ability to develop extensive and highly interconnected ERM networks, particularly if they are infective, could represent an important feature of efficient all-purpose AMF. However, the study of ERM in general has received little attention compared with the intraradical component of AM hyphae. This is due in no small way to the difficulties in carrying out such a study, particularly in the soil matrix, where there is a considerable problem of distinguishing the ERM from the mycelium of other fungi. Most of the initial studies were focused on hyphal density and viability and the usual methods employed for such estimations were destructive (Boddington et al., 1999) and more focused on the nutrient acquisition rather than the infective ability of ERM. Several methods have been used to study the structure and functioning of ERM (Frieze and Allen, 1991; Leak et al., 2004), including in vitro techniques (Bago et al., 1998). To make effective use of ERM as the preferential source of AM inoculum in agronomic systems, thereby promoting early colonization of crops, it is important to understand its persistence in soil as an infective propagule, the colonizing strategies of the AMF taxonomic cluster, the environmental conditions and, very importantly, the agronomic history of a particular site, particularly the cropping sequence of major host plants grown and the tillage system adopted.

### 6.3.1 Persistence of Infective Extraradical Mycelium in Soil

The turnover of ERM is considered to be quite fast, with hyphae of AMF living, on average, for 5–6 days (Staddon et al., 2003). Nevertheless, Olsson and Johnson (2005), using $^{13}C$ incorporation into signature fatty acids, found that most C assimilated by AMF remained for 32 days after labeling. Runner hyphae are a possible exception to the rule of fast turnover (Staddon et al., 2003) and the integrity of ERM and connection to plant roots is another key factor in maintaining and extending its ability to colonize new roots. Miller et al. (1995b) and Miller and Kling (2000) reported a much longer annual hyphal turnover of 26% in a prairie soil, where the development of ERM could be as great as 28 m $cm^{-3}$ soil. These discrepancies in reported turnover times are most likely associated with differences in sampling and methods of measurement.

ERM maintained infectivity overwinter under field conditions in eastern Canada with temperatures falling to -2°C, especially when soil remained undisturbed and in the presence of roots (Kabir et al., 1997b) or even without root connection (Addy et al., 1997). As long as there was a precooling phase, Addy et al. (1998) showed in an in vitro study that AMF hyphal activity was maintained after a period of 48 hours at -12°C. These are temperatures that are not commonly found in arable soils but highlight the ability of ERM to resist to cold temperatures. Therefore, cool temperatures do not seem to hamper ERM infectivity. ERM can also tolerate very warm temperatures, such as during the hot and dry summer conditions of the Mediterranean region, as described by Brito et al. (2011), where infectivity of intact ERM was maintained when air temperatures varied close to 30°C for almost 4 months with no rainfall and there was no surface vegetation.

The ERM is also resistant to water shortage. In undisturbed soil ERM was able to maintain its infectivity at very low (-21 MPa) matric potential and this was not dependent on it remaining attached to host plant roots (Jasper et al., 1989b). However, the capacity of hyphae of *Acaulospora laevis* to remain infective depended on the stage in the life cycle reached at the time of onset of the drying. If sporulation had commenced, then infectivity quickly declined as the soil dried. In contrast, infective hyphae of *Scutellospora calospora* survived for at least 11 weeks, regardless of the timing of the commencement of drying of the soil in relation to sporulation (Jasper et al., 1993). Nevertheless, wetting and drying cycles can have some negative impact on ERM infectivity (Braunberger et al., 1996).

Assuming that the results of these various studies reflect the response that occurs in the field under similar condition, it is reasonable to infer that an intact ERM, developed in the previous season, can persist through periods of temperature and water stress and be infective in the soil either in spring or fall, depending on the climatic zone, when seedlings emergence and root activity begins. This will therefore enable crop plants to become part of a functional mycorrhizal association earlier in their vegetation cycle and able to draw on the benefits of AM symbiosis sooner.

### 6.3.2 AMF Taxonomic Cluster Colonizing Strategies

The mechanisms involved in the establishment of a functional mycorrhiza may differ for species and genera of AMF (Dodd et al., 2000) but together with the large intraspecific functional diversity of mycelium growth (Munkvold et al., 2004) they influence ERM infectivity and AM effectiveness. Dodd et al. (2000) have expressed concern about the consequences of many studies having been limited to AMF species that are easy to maintain individually in open pot cultures and in vitro. One result is the limited knowledge about many other AMF groups. This is the case of the commonly described "fine endophytes," the Paraglomaceae, which are very difficult to isolate and maintain in open pot cultures (Abbott and Robson, 1982) but are frequently found when staining roots of plants from the field (Plate 6.3). Although molecular techniques are shedding some light on these groups, the difficulty of involving more diversity in comprehensive studies still persists.

From what has been described hitherto, it might be expected that AMF, which depend on spores for colonization, are more likely to be at a disadvantage compared with those able to colonize immediately from hyphal fragments or ERM. However, it is not that direct a relationship as there are several confounding factors, in particular the importance of the host plants and the negative feedbacks that occur within this mutualistic symbiosis (Bever, 2002). Some studies indicate that ERM is highly infective in the Glomaceae and Acaulosporaceae whereas members of the Gigasporaceae regenerate mostly from spores (Tommerup and Abbott, 1981; Morton, 1993), although these conclusions are only partially supported by later studies. According to Hart and Reader (2002a) AMF variation of colonizing strategies is taxonomically based at the family level and from the 21 isolates tested, they observed that most Glomaceae isolates colonized roots before

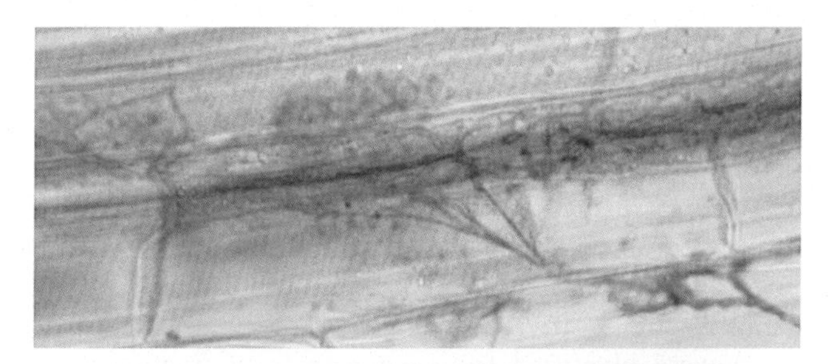

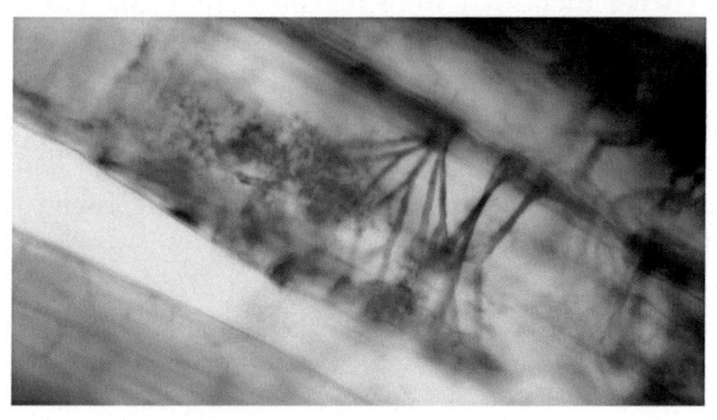

**PLATE 6.3** Wheat roots colonized with "fine endophytes" (200×).

Acaulosporaceae and Gigasporaceae isolates, and produced a high level of root colonization but a low level of soil colonization, while Gigasporaceae showed an opposite trend and Acaulosporaceae isolates resulted in low levels of root and soil colonization. Klironomos and Hart (2002), studying eight AMF isolates and a single host plant, found that *Glomus* and *Acaulospora* isolates colonized from all inoculum types, whereas *Gigaspora* and *Scutellospora* isolates colonized mainly from spores and to a limited degree from root fragments, and ERM was not an effective propagule of the tested species of *Gigaspora* and *Scutellospora*. Drew et al. (2006) theorized that some AMF fungal species, like *Glomus mosseae*, may produce large amounts of external hyphae primarily to increase the probability of locating and colonizing a new host plant. This also may result in an initial negative growth response of the host plant following colonization.

In combining the evidence from Brito et al. (2012b) and Brígido et al. (2017), it appears that even if ERM does not seem to be the most suitable propagule for species of *Gigaspora* and *Scutellospora*, they are able to persist and colonize different host plants in undisturbed soil systems (no-till) (Jansa et al., 2003), where ERM is considered to be the main inoculum source. It therefore appears that different colonization strategies and AMF biodiversity can be maintained under these tillage systems. In contrast highly disturbed systems that rely on inversion tillage tend to unbalance AMF populations, favoring colonizers like *Glomus* spp. that readily infect host plants from spores (Dodd et al., 2000) (see Section 4.4).

Comprehensive field-level studies on AMF diversity among the different propagule types are scarce but now rely on the more robust new generation based identification methods. Varela-Cervero et al. (2015) carried out the first research to investigate the explicit diversity of AMF with respect to soil spores, ERM and AMF root structures associated with five Mediterranean plant species in a natural park (Sierra de Baza, Granada, Spain) at a single sampling time in spring. The authors concluded that AMF communities varied significantly among propagule fractions and the root-colonizing fractions differed among host plant species. Some 71 virtual taxa (VT) were identified from the plant root fraction and 47 from each of the spore and ERM fractions. Roots and ERM had more VT and representatives of those taxa in common than did roots and spores, consistent with the better colonization ability of ERM. Gigasporacea VT were the most abundant group and were mainly found in the ERM and spore fractions with very few representatives of this taxa being found in the roots, which supported earlier culture-based studies.

Following that study Varela-Cervero et al. (2016a) confirmed different colonization strategies for the different AMF families involved. Glomeraceae and Claroideoglomeraceae initiated colonization mainly from colonized roots, whereas for Pacisporaceae and Diversisporaceae it was from spores and ERM. The authors also found that AMF community composition in roots differed significantly between spring and autumn; however, no significant differences were detected in soil propagules, spores and ERM, between the two seasons (Varela-Cervero et al., 2016b). This latter study also highlighted the central role of the host plant in the colonization process.

### 6.3.3 The Presence of Adequate Host Plants and ERM Integrity: Crop Rotations and Tillage Regime

Hart and Reader (2005) developed a quantitative study to investigate the role of ERM in early AM colonization for three AMF isolates (*Glomus etunicatum, G. intraradices,* and *Gi. gigantea*), which represented a continuum of colonization strategies. This choice of isolates allowed Hart and Reader (2005) to determine if the AMF with the most rapid colonization would have more infective hyphae and contact points. They concluded that extensive root colonizers (*G. intraradices* and *G. etunicatum*) produced more but later infection units than did species with limited root colonization (*Gi. gigantea*). The two *Glomus* species also produced more external structures early on. Apparently, an extensive soil colonizer like *Gi. gigantea* does not necessarily produce more external structures in general, including runner hyphae, hyphal bridges and, in particular, more extensive absorptive hyphae than AMF with a limited soil mycelium.

These findings suggest that to evaluate the potential of the ERM as inoculum source at a particular site, the colonization rate of a given mycotrophic host plant can be indicative of the infective ERM in the soil available to colonize the following host plant. Brito et al. (2014) and Alho et al. (2015) studied the colonization of wheat and clover by a native AMF population following different ERM Developer plants. Greater AM colonization of wheat and clover was observed when these plants were grown after mycotrophic host developer plants (*O. compressus* or *Lolium rigidum*), than after the scarcely mycotrophic host plant (*R. bucephalophorus*), when the soil was kept undisturbed (Table 6.1). However, if the soil was disturbed and the ERM disrupted, the beneficial effect of the previous plant on the development of the ERM network was completely lost, and the initial colonization rates of wheat or clover declined drastically to levels observed when the previous plant was the scarcely mycotrophic host *R. bucephalophorus*. The effect of an early AM colonization, promoted by the intact ERM, was already observed in wheat, only 10 days after planting (DAP), when it increased from 0.4% after *R. bucephalophorus* to 5.8% after *L. rigidum*, and 7.3% after *O. compressus*. These differences were still evident 21 DAP and were reflected in a greater shoot dry weight and P content for both crops (Table 6.1). In the case of clover a better rhizobium performance was also recorded from 21 DAP, and although by 42 DAP the initial differences of clover arbuscular colonization between treatments disappeared, greater shoot, and nodule dry weight persisted (Section 5.1).

**TABLE 6.1** Arbuscular colonization (AC) of host plants used to develop extraradical mycelium (ERM) and AC of wheat and clover grown for 21 days after the first plants in undisturbed soil (ERM intact) and disturbed soil (ERM disrupted).

| | | ***Rumex bucephalophorus*** | | | ***Ornithopus compressus*** | | | ***Lolium rigidum*** | | |
|---|---|---|---|---|---|---|---|---|---|---|
| | | **AC** | **SDW** | **P** | **AC** | **SDW** | **P** | **AC** | **SDW** | **P** |
| Developer plant | | 2 c | – | – | 74 a | – | – | 52 b | – | – |
| **Plant after developer** | | | | | | | | | | |
| Wheat | ERM Intact | 17.5 cd | 99 c | 0.1 c | 56.4 a | 254 a | 0.6 a | 49.8 a | 158 b | 0.4 b |
| | ERM Disrupted | 8.8 d | 80 c | 0.1 c | 13.5 cd | 93 c | 0.1 c | 18.2 bc | 84 c | 0.1 c |
| Clover | ERM Intact | 33 bc | 36 c | 1.0 b | 68 a | 104 a | 2.8 a | 62 a | 64 b | 2.5 a |
| | ERM Disrupted | 22 c | 34 c | 0.7 b | 24 c | 32 c | 0.6 b | 40 b | 38 c | 1.0 b |

Values with the same letter within wheat, clover, and Developer plants are not significantly different at $P = .05$.
Arbuscular Colonization (%); SDW, Shoot Dry Weight (mg plant$^{-1}$); P, Shoot Phosphorus Content (mg plant$^{-1}$).
Adapted from Brito et al. (2014) and Alho et al. (2015).

Results present in Table 6.1 clearly establish the importance of the host plant in the development of an infective ERM in the soil and the need to keep it intact to ensure the ability to effect early colonization of the subsequent plants. In an agronomic context both factors are closely connected to the plants present in a cropping system – rotation, cover crops, weeds – and the adoption of tillage techniques, such as no-till or conservation tillage, that cause reduced disturbance of the soil (see Sections 2.1 and 2.2). Therefore, the cropping history of a particular site is particularly relevant if we want to promote ERM as the main source of AMF propagules to favor an early establishment of an AM with the crop and capitalize sooner on its benefits.

The inclusion of nonmycotrophic crops such as sugar beet and canola in a rotation leads to a general reduction in the number of mycorrhizal propagules. On the other hand, the cultivation of mycorrhizal host crops increases AMF populations and maintains mycorrhizal activity in soil (Arihara and Karasawa, 2000; Gollner et al., 2004; Vestberg et al., 2005). Reduction of AM fungal propagules can also be a significant consequence of adopting a bare-fallow due to the absence of host plants (Kabir and Koide, 2002) (see Section 4.4). The so-called long fallow disorder is caused by a decline in viable propagules of AMF during the absence of host plants, resulting in poor root colonization and symbiotic effectiveness in a subsequent crop (Thompson, 1987). Apparently a bare-fallow can be even more detrimental to AMF propagules than the presence of a nonmycotrophic crop in the rotation. For example, Harinikumar and Bagyraj (1988) reported a 40% and 13% decrease in AMF propagules after fallow and a nonmycorrhizal crop, respectively.

As generally described in the preceding paragraphs, the integrity of ERM is essential for its efficacy as the propagule for earliest AM formation. The implements used for soil preparation (mechanical weed control and seedbed formation) can easily disrupt the ERM, particularly with inversion tillage, where ploughing fragments ERM and inverts soil layers. Furthermore other operations usually associated with seedbed preparation, scarification and disk harrowing, completes the final disruption of ERM. Ploughing to more than 15 cm hinders subsequent mycorrhiza formation by reducing propagule density in the upper part of the rooting zone (Kabir et al., 1998; Abbott and Robson, 1991). No-till or reduced tillage techniques are therefore much less damaging to ERM as the soil is moved only in the shallow layers where the seed is placed, keeping intact ERM in close proximity to the seed, thereby increasing the likelihood that a newly emerging root will intersect an AMF propagule capable of prompt colonization. In addition, in no-till systems new roots follow the old root channels that are maintained and also can encounter more AMF propagules than plants growing in soil that has been tilled (Kabir, 2005). Both aspects of the reduced soil disturbance could be expected to allow faster and greater AMF colonization of plants cultivated under no-till or reduced tillage than under inversion tillage systems (Kabir et al., 1997a; McGonigle and Miller, 1996; Kabir et al., 1998; Galvez et al., 2001; Castillo et al., 2006a;

Brito et al., 2012a). Nevertheless large inoculum density in ecosystems, such as the Australian pasture studied by Jasper et al. (1991) and discussed by McGonigle and Miller (2000), can likely override any effect of soil disturbance and ensure roots of all plants become well colonized by AMF.

## 6.4 MULTIPLE ROLES OF ERM AND COMMON MYCORRHIZAL NETWORKS

*The ERM provides a superior propagule source for AMF and aids acquisition of soil nutrients. However it contributes to many other important functions at soil and plant level.*

We have stressed the importance of an intact ERM as the preferential propagule source for early development of AM in host plants but its multiple roles in soil are far more extensive. ERM hyphae are thinner than roots or roots hairs (see Plate 3.1) and can occupy soil pores inaccessible to the latter and extend over considerable distances. In agricultural field soil up to 50 m of AM hyphae per gram of soil have been observed (Smith and Read, 2006) and the ratio of fungal hyphae to roots in soil can be 100:1 or even greater in natural ecosystems (George et al., 1995). The most prominent role of ERM is to take up nutrients from the soil, transfer these nutrients to the intraradical mycelium within the host root, and exchange the nutrients for carbon from the host across a specialized plant-fungal interface (arbuscules) (see Section 3.1.1 and Box 5.1). Between 20% and 45% of plant photosynthetic activity may be involved in the exchange (Smith and Read, 2006).

In general AMF exhibit little host specificity and the same AMF can colonize concurrently various host plants, even from different species, and establish connections between them (Heap and Newman, 1980). Hence a single individual may form a common mycorrhizal network (CMN) between several coexisting plant individuals from different species (Walder et al., 2012). These networks can potentially be indefinitely enlarged by hyphal fusions since anastomosis between hyphae originating from the same spore and from different spores of the same isolate of AMF have been described (Giovannetti et al., 1999, 2004). After several years of research, initially in microcosms and more recently in natural ecosystems, CMN are considered to be major components of terrestrial ecosystems (Selosse et al., 2006).

### 6.4.1 Transfer of Nutrients Between Plants

One important consequence of CMN is the transfer of nutrients between plants. A large ERM as part of a network can effectively absorb nutrients from soil and also from dying roots and thereby represent a readily available pool of nutrients for the interconnected plants (Heap and Newman, 1980). Determination of nutrient transfer via CMN between living plants is more complex. Even so, Whittingham and Read (1982) reported $^{32}P$ transfer between 3 months old (mature "source") and 3 weeks old (younger "sink") plants of *Festuca ovina* and similar results were observed with *Plantago lanceolata*. Mikkelsen et al. (2008) also described $^{32}P$ transport toward the largest sink of living donor and receiver plants, over a distance of at least 20 cm, via an anastomosed mycorrhizal network.

Numerous studies over the past three decades have demonstrated that plant-to-plant N transfer from the donor to the receiver can take place directly through mycorrhizal hyphae in CMN interconnecting roots (He et al., 2009). Transfer of $NH_4^+$ or $NO_3^-$ between $N_2$-fixing plants and non-$N_2$-fixing plants mediated by CMN have been described (van Kessel et al., 1985; Johansen and Jensen, 1996; Moyer-Henry et al., 2006), even if competition between multiple potential receivers can occur (Walder et al., 2012). As in the case of P, unidirectional N transfer via CMN occurs toward the largest sink, and the driving force seems to be N gradients between N-rich donors and N-limited receivers (Bethlenfalvay et al., 1991; Frey and Schüepp, 1992). However, the $N_2$-fixing plant is not always necessarily the N donor, as in few cases N transfer took place through a CMN from non-$N_2$-fixing plants to $N_2$-fixing plants, with net gains for the recipient plant (Johansen and Jensen, 1996; He et al., 2009).

Whether or not there is carbon transfer via CMN between established plants and from established plants to seedlings is a matter of debate and can be related to plant-fungus combinations and experimental conditions. After an extensive review, Jakobsen and Hammer (2014) state that C transfer at the symbiotic interface occurs in the direction of the fungal nutrient source, while mineral nutrient transfer occur in the opposite direction toward the plant C source. Voets et al. (2008) provided clear evidence that carbon was transferred from a donor to a recipient plant via the CMN network, using two photosynthetically active plants. However, the transferred carbon remained within the intraradical AMF structures of the recipient plant root and was not transferred to the plant tissues.

For Jakobsen and Hammer (2014), instead of being considered as a system for interplant nutrient transfer, the CMN rather represents the mechanism for potential resource acquisition from a shared pool. The ERM may on one hand obtain C from several individual plants, while on the other hand it serves as an efficient mesh for absorbing mineral nutrients to cover its own need and to distribute the surplus among the plant components of the CMN. The regulatory role of nutrient distribution in a shared mycelium network and its impact on plant competition and seedling establishment remains to be resolved and plants that constitute the greater sink will benefit most from the CMN in this context. The identity of plant and fungal partners together with prevailing environmental conditions, most probably play an important role in the nutrient trade within the network.

Nevertheless the shortcut allowed by the CMN avoids nutrient retention at soil level and helps to maintain a tight nutrient cycle, keeping nutrients more directly available to the plants (Mikkelsen et al., 2008). In agricultural systems, during seedling establishment, this function of the CMN may be a most relevant aspect as it provides not only direct access to a large pool of soil nutrients but also direct access to nutrients from the residues of the previous crop or decaying weeds.

### 6.4.2 Communication Between Plants

Interplant signaling can also occur through CMN (Barto et al., 2012) and influence plant defense mechanisms against foliar necrotrophic fungi and insect herbivores by the direct transfer of signaling molecules (Song et al., 2010, 2014; Babikova et al., 2013). Induced defense signals can be transferred between the healthy and pathogen-infected neighboring plants, suggesting that plants can perceive the defense signals from the pathogen-challenged neighbors through the CMN and activate defenses before being attacked themselves. According to Johnson and Gilbert (2015), interplant signaling via CMN offers a potentially more effective and directed pathway of defense compared with the aerial pathway, because signals may be able to travel greater distances, and are unaffected by dilution effects caused by wind.

CMN can also have some deleterious effect on plant growth since they can have a significant role in the transfer of allelochemicals (Achatz and Rillig, 2014), increasing the effectiveness of the transfer in natural systems and interfering in chemical interaction processes in the soil.

### 6.4.3 Development of Soil Structure

Mycorrhizal fungi and their ERM in particular, make direct contributions to soil aggregation and aggregate stability (Tisdall and Oades, 1982). This role can be enhanced in no-till systems where hyphal networks remain intact. The direct effect of ERM on soil aggregate formation, by the enmeshing of soil particles, was shown by Thomas et al. (1993) to be significant and at least equivalent to that of roots alone, primarily at macroaggregate level. Glomalin and glomalin-related soil proteins (Wright and Upadhyaya, 1996; Rillig, 2004b), are specific glycoproteins produced by AMF that have been claimed to make important contributions to soil aggregate stability as cementing agents (Miller and Jastrow, 1990), especially at the microaggregate level. These agents are located on the surface of active AM hyphae (Rillig et al., 2001) and tightly bound within the hyphal wall of AM fungi (Driver et al., 2005) playing a role in the living fungus, and its functionality in the soil appears only to be secondary, possibly due to its relatively slow turnover rate in the environment (Steinberg and Rillig, 2003; Driver et al., 2005; Goss and Kay, 2005; Rillig and Mummey, 2006).

The amount of glomalin-related soil proteins and water stable aggregates is highly correlated and greater in fields under no-tillage, where the ERM is kept intact (Curaqueo et al., 2010). However, the relationship reported between aggregate stability and glomalin content varies greatly between different plant species on the same soil (Goss and Kay, 2005), which indicates that other processes are also involved in stabilization. Once again the specificity of AMF isolates and plant host is an important factor and the effect of AM on soil aggregation depends on the interaction between the symbionts (Bedini et al., 2009; Piotrowski et al., 2004).

There are numerous microbes associated with ERM, such as bacteria, that produce a variety of external polymeric compounds, which can have an important influence on soil aggregation (Rillig et al., 2005; Albertsen et al., 2006) and the stability of soil structure in general, some by lining soil pores. Also among these accompanying microbes are plant growth promoting rhizobacteria, which find an important survival niche where physical, metabolic and functional complex interactions take place (Larsen et al., 2009). Synergistic interactions have been described between these bacteria and AMF in enhancing plant growth under normal (nonstress) or stressed environments (Nadeem et al., 2014). Such microbes may be vital for sustainable agriculture because their functioning depends on biological processes to maintain plant growth and development under stress conditions, further contributing to the enhancement of soil ecosystem services.

## 6.5 CONCLUSIONS

In the framework of an agronomic system, all the intricate variables governing AM symbiosis make it impossible a priori to identify a perfect AMF × host plant × site combination and generalized solutions. Even so there are crucial aspects to consider:

- ERM integrity is critical in making effective use of the opportunities AM mycorrhiza offer. Tillage systems that need to be adopted require no-till or reduced tillage techniques as opposed to inversion tillage techniques.
- An intact ERM as the preferential AMF propagule appears to be essential to ensure an earlier and faster AM establishment in crops.
- Intact ERM has a considerable ability to remain infective, even in less favorable environmental conditions (temperature and water availability) and over long periods, allowing its management and exploitation as preferential AM inoculum source between cropping seasons, irrespectively of agri-climatic region.
- The design of appropriate crop rotations and components, including optimizing the presence of mycotrophic plants and avoiding long bare-fallow periods, is fundamental to ensure the persistence of ERM in the soil. These objectives also help to guarantee AMF biological and functional

diversity and maintain the greatest potential for successful AMF × host plant combinations.
- CMN offer a broad range of favorable possibilities that were little known until recently.

In practical terms all these aspects can be applied in the field and managed within agronomic practices to improve production through the multiple benefits of the AM symbiosis. These insights are further explored in Chapter 8.

Chapter 7

# New Tools to Investigate Biological Diversity and Functional Consequences*

Chapter Outline



If we are to exploit the opportunities offered by our increased understanding of arbuscular mycorrhiza (AM) diversity and the potential to manage it, we need greater knowledge of the indigenous arbuscular mycorrhizal fungi (AMF) that are participating in symbiosis with the target plants. In this chapter we consider the role of new techniques available for a taxonomic description of the AMF community present in the soil and host plant and their quantitative and qualitative contributions to improving the ability of that host to grow in its environment. Ideally we would like to know how many species or taxonomic groups are present, the number of individuals that can be identified as belonging to the same group, and how much variability there is within such a group in terms of their capability to enhance growth. With the techniques available we can be precise over the molecular makeup of individuals but it is more difficult to be certain as to which species they may belong to, particularly if that species has been mainly identified through the morphological characteristics of its spores, rather than by DNA sequencing. Consequently we often refer to an *operational taxonomic unit* (OTU) or *virtual taxa* (VT) (Öpik et al., 2014) to describe a taxonomic group identified solely through molecular techniques. Indeed, it has been suggested that the concept of *species* is inappropriate for AMF and that the term AMF *taxa* would be better (van der Heijden et al., 2004).

* With Diederik van Tuinen.

**Functional Diversity of Mycorrhiza and Sustainable Agriculture.**
**DOI: http://dx.doi.org/10.1016/B978-0-12-804244-1.00007-1**

The number of different OTUs of AMF present in the soil or within the assemblage forming an AM with a particular host provides one level of assessment of the biological diversity in the AMF present. For assemblages of AMF, the ability of different OTUs to modify the level of response of a host plant to the presence of a biotic or abiotic stress provides one assessment of the functional diversity.

## 7.1 GENETIC MARKERS

A *genetic marker* is a gene or DNA sequence that can be used for the identification or detection of a specific cell or organism. In the laboratory, it is possible to make multiple copies of a particular segment of DNA (*amplification*) if a short fraction of the sequences on both sides bordering a region of interest (primers) are known. The identification of individuals depends on the specific differences (*polymorphism*) in the nucleotide sequence amplified, or in the primer sequence, that have developed during the evolution of the ancestors of each individual AMF and therefore help establish its taxonomic affinity (phylogenetics) and identity (*barcoding*).

*Within AMF hyphae there are cytoplasmic particles, called ribosomes, where genetic information from the DNA is used to construct proteins essential for the organism to function. The DNA also carries the information necessary for the formation of a ribosome. Each ribosome is composed of RNA and protein and is formed of two subunits, one being about half as big again as the smaller subunit.*

### 7.1.1 Small Ribosomal Subunit

Before the technical revolution in biological research, brought by the development of the *polymerase chain reaction* (PCR) (Mullis et al., 1986) allowing the amplification of specific regions of the genome, the method of choice for molecular taxonomy was the use of DNA:rRNA hybridization, using 16S RNA as probe (De Ley and De Smedt, 1978; Doolittle et al., 1975). This method was mainly applied to bacterial taxonomy and was further improved with the sequencing of the rRNA molecule and mainly associated with the small ribosomal subunit (SSU) or 16S gene (Lane et al., 1985; Qu et al., 1983). The reason for this choice was that nearly 95% of total cellular RNA is ribosomal RNA (rRNA) and made of an equimolar relation between the SSU and large ribosomal subunit (LSU) RNA. For practical reasons the entire sequence of the SSU rRNA was easier to obtain than that for the LSU rRNA, which for bacteria are comprised of 1500 and 2900 nucleotides, respectively (Woese, 1987). Therefore when PCR technology became available and isolation of large amounts of DNA was no longer required, a large dataset of SSU sequences, covering a wide range of phylum from bacteria to human, was already available. This region was therefore chosen for the first investigation of arbuscular mycorrhizal diversity (Simon et al., 1993) with the generation of the first PCR primer claimed to be specific for AMF DNA: VANS1 (Simon et al., 1992). However, as the amount of information has increased this VANS1 primer has been shown to lack specificity for the *Glomerales* and several fungi of the Glomeromycota cannot be amplified with this primer (Schüßler et al., 2001a). To solve this problem the eukaryotic specific primer NS31 (Simon et al., 1992) was used in combination with a new primer AM1 (Helgason et al., 1998) but the latter failed to amplify the rDNA from the ancient Glomeromycota linages of *Archeosporales* and *Paraglomerales* (Redecker et al., 2000; Schüßler et al., 2001a) and can also amplify some sequences from fungi that are not members of the Glomeromycota (Douhan et al., 2005; Rodriguez-Echeverria and Freitas, 2006; Santos-Gonzalez et al., 2007). New sets of primers (AML1 and 2) covering a larger region (~800 kb) of the SSU were then further developed (Lee et al., 2008). Another region, the internal transcribed spacer (ITS), has been used for selecting primers specific to the Glomeromycota, with the development of group specific primers covering the different genera (Redecker, 2000), and has recently been chosen as the universal barcode maker for *fungi* (Schoch et al., 2012), although other barcoding markers are more adapted to the Glomeromycota (Schoch et al., 2012).

### 7.1.2 Large Ribosomal Subunit

The LSU gene region has also been used in the molecular analysis of Glomeromycota diversity, taking advantage of the higher level of discrimination that this region has relative to that of the SSU (Schoch et al., 2012; Van de Peer et al., 1996). The molecular polymorphism of the 5′-end of the LSU gene, mainly in the D1 and D2 domains (Liu et al., 2012a; van Tuinen et al., 1998a), allowed discrimination between taxa (van Tuinen et al., 1998a; Jansa et al., 2008), taxon-specific primers at the species level (van Tuinen et al., 1998b; S. Trouvelot et al., 1999; Turnau et al., 2001; Gamper and Leuchtmann, 2007; Pivato et al., 2007) and at a higher level such as genera (Rosendahl and Stukenbrock, 2004) or phylum (Gollotte et al., 2004). These primers have mainly been used in "end point" PCR for diversity studies but have also been applied to following mycorrhiza inoculation of a field (Farmer et al., 2007) and for studies using quantitative PCR (qPCR), where the proportion of each different AMF is determined (Alkan et al., 2006; Pivato et al., 2007; Gamper et al., 2008; Jansa et al., 2008).

The choice of primers covering an entire phylum is always problematic and often a compromise between a positive bias toward some genera, or lack of specificity with the amplification of non-Glomeromycota sequences

(Kohout et al., 2014). For example, the primer FLR4 (Gollotte et al., 2004), which has specificity at the phylum level, has been used in various studies (Lekberg et al., 2013; Matekwor Ahulu et al., 2006; Mummey and Rillig, 2007; Pivato et al., 2007; Holland et al., 2014). However, it does not match all Glomeromycota fungi perfectly (Gamper et al., 2009; Krüger et al., 2009) but the Glomeromycota families amplified are those found mainly in agro-ecosystems (Bouffaud et al., 2015). To overcome this limitation Krüger et al. (2009) proposed the use of a combination of a several forward and reverse primers, used in a nested-PCR approach to cover the 3′-end of the SSU and the 5′-end of the LSU, amplifying a region of 1500–1800 nucleotides, respectively, depending on the particular species of Glomeromycota (Krüger et al., 2009; Schüßler et al., 2015; Senés-Guerrero and Schüßler, 2016). The aim of using a combination of several primers at the same time is to increase the number of families with a perfect match, although keeping the number of primers to a minimum. The region amplified covers less variable regions such as the 5.8S, but also some highly polymorphic regions such as ITS2, D1, and mainly the D2 domain of the 5′-end of the LSU. The LSU D2 domain is the ribosomal gene region with the highest level of polymorphism (Michot and Bachellerie, 1987; van Tuinen et al., 1998a; Kohout et al., 2014). In the Glomeromycota there is an important length polymorphism. For example, the D2 domain of the *Glomerales* differs from that of the *Diversisporales* by about 50 nucleotides (van Tuinen et al., 1998a). Recently, however, the primers developed by Krüger et al. (2009), have also been shown to amplify non-Glomeromycota sequences under some conditions (Kohout et al., 2014; Van Geel et al., 2014), especially when the DNA is extracted from soil (Krüger et al., 2015), and thereby generate a significant fraction of chimeric sequences (Kohout et al., 2014).

*The development of molecular methods for identifying the AMF present in a population has required the development of an approach to establish what constitutes a taxonomic unit.*

### 7.1.3 Delineation of Operational Taxonomic Units

Glomeromycota fungi contain several hundreds of nuclei that show a low level of intersporal (Tisserant et al., 2013) or internuclei polymorphism (Lin et al., 2014). However, the nuclear ribosomal encoding gene, in contrast to the rest of the genome, is highly diverse (Sanders et al., 1995; Lanfranco et al., 1999; Pawlowska and Taylor, 2004). The consequence is that several ribosomal sequences are obtained even when starting from the same spore, implying that divergent sequences can be obtained from the same species. Therefore the assignment of a ribosomal sequence to specific taxa, requires the use of a clustering process with an arbitrary threshold (Lekberg et al., 2014; Powell and Sikes, 2014). The use of universal OTU delineation (97%) (Powell and Sikes, 2014) has been shown to generate similar diversity data to that of an evolutionary based process (Lekberg et al., 2014). To help in the assignment of OTU obtained after the clustering processes, the use of reference sequences is essential. Several curated fungal databases have been set-up that specifically cover the fungal ITS region; one such is the UNITE database (Kõljalg et al., 2013). The SILVA database (Quast et al., 2012) covers the fungal ITS and includes the two ribosomal subunits. More recently the MaarJam database (Öpik et al., 2010) was created and specifically dedicated to data on the Glomeromycota. Originally it contained only sequence data for SSU rRNA but, with the availability of other sequence information, this database now includes the ITS and LSU ribosomal sequence data. Currently MaarJam also contains several sequences that code for protein such as actin, EF1-alpha, RPB1, or mitochondrial large ribosomal subunit (mtLSU) (Öpik et al., 2014). Such a database is essential for cross-validation of an OTU and it will aid comparisons between studies and ease the understanding of AMF ecology and biogeography, even though a large number of the newly obtained sequences are related to uncultured Glomerales, consistent with the view that hitherto mycorrhizal diversity seems to have been highly underestimated (Hibbett, 2016).

*In fungal hyphae, mitochondria are organelles where the sugar from the host plant is converted into energy. Mitochondria carry their own DNA and also form ribosomes for the production of the proteins required for their maintenance and activity. These ribosomes are smaller than those formed by the nuclei of the fungus.*

### 7.1.4 Mitochondrial Large Ribosomal Subunit

Other genetic markers, such as the mtLSU sequence, have been used to bypass the intranuclear ribosomal polymorphism. The mtLSU region markers were studied (Börstler et al., 2008; 2010) and used with success (Börstler et al., 2008; 2010), but difficulties arose owing to the large length variation observed, even between closely related species. For example, within the genus *Rhizophagus* the mtLSU varies from 1070 to 3935 base pairs. This important variation in the sequence length can be advantageous for the generation of isolate-specific primers, but is not adapted to community analysis by next generation sequencing techniques (NGS). However, this genetic marker is useful when used in combination with a Restriction Fragment Length Polymorphism (RFLP) approach (Peyret-Guzzon et al., 2015).

### 7.1.5 RNA Polymerase II

Improvement in the resolution power of phylogenetic studies requires several genomic markers to be used. For example, to establish a phylogenetic relationship for the yeast genus *Saccharomyces* with a confidence of 95%, Rokas et al. (2003) showed that a total of eight concatenated genes were required. For most organisms, information on such a large number of marker genes are not available but multiple gene phylogeny has already been used for fungi (James et al., 2006). The limited number of marker genes available for investigating the Glomeromycota severely limits this approach and more information is required to improve the phylogenetic study of AMF. The large subunit of the RNA polymerase II (RPB1) gene has been shown to be an interesting marker for phylogenetic analysis as it is present in single copy, with no intraisolate variants (Tisserant et al., 2013), and with the presence of *introns* (regions of the genome that do not code for proteins) that are rich in polymorphism (Stockinger et al., 2014). This marker has been used for diversity studies (Stockinger et al., 2014; Peyret-Guzzon et al., 2015) but the number of primers required to cover the whole Glomeromycota phylum is challenging for NGS approaches.

> *Whereas biological diversity within AMF relates solely to genotype, functional diversity also depends on the host plant and the wider environment. It is the variation expressed in terms of the impact on the host plant in the presence of stresses.*

## 7.2 FUNCTIONAL DIVERSITY

Hart and Reader (2002a) investigated the possibility of AMF colonization strategy varying between major taxonomic groups based on mycelium characteristics. They selected 12 members of the family Glomaceae, 5 members of the Gigasporaceae, and 4 members of the Acaulosporaceae and made an inoculum comprising spores, colonized root fragments, and hypha for each. Each inoculum was applied to narrowleaf plantain (*Plantago lanceolata* L.), common plantain (*Plantago major* L.), Kentucky-blue grass (*Poa pratensis* L.), and annual blue grass (*Poa annua* L.) grown in a controlled environment. The members of the Glomaceae were the earliest to colonize the hosts, probably from the colonized root fragments, and produced the most mycelium but the majority of this was inside the roots rather than the soil (Table 7.1). This contrasted with the mycelium of the Gigasporaceae, which was mainly extraradical and had a much greater hyphal density in the soil than did the members of the other families. Members of the Acaulosporaceae produced a lot of mycelium but were slow to colonize the host plants (Table 7.1), likely because they mainly colonized from spores, as did the members of the Gigasporaceae.

**TABLE 7.1** Characteristics of the mycelium of the members of three families of arbuscular mycorrhizal fungi (AMF), including speed of colonization
(Source: Data from Hart and Reader, 2002a)

| | Glomaceae | Gigasporaceae | Acaulosporaceae |
|---|---|---|---|
| Time for hosts to be colonized (weeks) | ~4 | 6–8 | 6–8 |
| Proportion of intraradical to extraradical mycelium | 5.3 | 0.4 | 2.4 |
| Range of % root colonized after 70 days | 21–85 | 5–73 | 5–73 |
| Hyphal density in soil ($m\ cm^{-3}$) | 1–2 | 6–9 | 1–2 |

In a second experiment, Hart and Reader (2002b) evaluated the impact of the differences in mycelium characteristics on the growth of the host plants and the P content of their leaves. The host plants benefitted most from colonization by AMF having greater intraradical mycelium. In this case members of the Glomaceae produced greater increases in biomass of the three host plants than did the representatives of other families. Some members of the Acaulosporaceae produced the greatest dry matter increase when the host plant was *Poa annua*. Across all host plants the increase in biomass was not directly related to the P content of the leaves, and there was considerable variability between genera and isolates in their effects on the different host plants. The results suggested that although AMF family traits were important in the formation of effective mycorrhiza, host-fungus specificity and the variability within families were also significant factors.

Maherali and Klironomos (2007) again used isolates from these same AMF families having similar mycelium characteristics to those described in Table 7.1. Using narrowleaf plantain as host plant, a series of treatments consisting of either isolates from a singular family, two families or all three families. A total of eight isolates were used in each treatment but only four different isolates were available for the Acaulosporaceae. After a year the number of isolates remaining was 2.7, when only a single family was applied. When two families (either Glomaceae or Gigasporaceae) had been applied, an average of 5.4 isolates remained after 1 year, with slightly fewer members of the Gigasporaceae surviving than Glomaceae. The average was 6.8, when three families had been applied, with slightly more members of the Gigasporaceae surviving than Glomaceae. Plant biomass was greatest when at least two families were involved in the assemblage of AMF.

In this experiment the greater ERM of the Gigasporaceae was associated with more P in the leaves of narrowleaf plantain than with the greater intraradical mycelium of the Glomaceae and the Acaulosporaceae. At the same time the members of the Glomaceae were more effective in reducing the level of root colonization by two root pathogens, *Fusarium oxysporum* and a *Pythium* species, than were members of the other two families. Overall the successful assemblages combined the ability to acquire P and to protect the host plant against pathogens, consistent with greater diversity being associated with greater ecosystem functioning through enhanced trait richness (Maherali and Klironomos, 2007).

Considerable intraspecific variability in function has been shown for P acquisition, mainly through the ability to exploit nutrient patches through differences in hyphal growth patterns (Munkvold et al., 2004). This is consistent with the concept that functional complementarity exists within AMF and can enhance ecosystem functioning but specifically supports the view that it can occur between AMF isolates within the same species. It suggests that functional diversity and species diversity are not necessarily related (van der Heijden et al., 2004). Nevertheless, many studies have shown that the above and belowground diversity are related (van der Heijden et al., 1998b) and in a positive way, with the larger the belowground diversity the greater the beneficial effect on the plants (van der Heijden et al., 1998a), even when the level of mycorrhizal dependency varies between plant species. Yet the link between diversity and functionality is still poorly understood, in part because the techniques available are limited. The most commonly used methods to assess the status of the mycorrhizal colonization are based on the estimation of the extent of fungal presence revealed after root staining (Philipps and Hayman, 1970; Trouvelot et al., 1986; McGonigle et al., 1990). These methods estimate the morphological development of the fungus in the root system as well as the number of arbuscules, which are the principal structures for nutrient exchange between the plant and the fungus. Although widely used the methods provide no link between the intensity of root colonization and the amount of nucleic acid in the fungus that is encoding ribosomal genes (Pivato et al., 2007) or the functional significance of fungal activity (Tian et al., 2013b). Quantification of a particular fungal isolate, even if identification or enumeration can achieve this level of resolution, cannot yet be related to a specific function. Mycorrhizal symbiosis deeply affects the physiology and affects an important number of metabolic functions in the host plant (see Chapters 2, 3 and 5). Mycorrhiza, for example, improve the uptake of nutrients such as phosphate (Maldonado-Mendoza et al., 2001; Smith et al., 2011), nitrogen (Hodge et al., 2001; Cappellazzo et al., 2008), or microelements, such as copper or zinc (Smith and Read, 2008), but can protect the plant from adverse effects of toxic levels of metals and metalloids (Brito et al., 2014; Alho et al., 2015). Approaches based on protein-encoding genes has been used with considerable success in bacterial ecology, targeting genes involved in specific biochemical processes, such as $N_2O$ reductase by the denitrifier community (Philippot et al., 2009), cycling of nitrogen or carbon (Kowalchuk et al., 2006). For each of these physiologically based functions, specific gene markers have to be identified and monitored (see Box 7.1). These studies targeted not a specific group of organisms but key genes of specific biological pathway with the use of degenerated primers due to the genetic code degeneration (Rotthauwe et al., 1997; Philippot et al., 2009). Although all the fungi were able to establish an arbuscular mycorrhizal symbiosis belong to a single phylum, Glomeromycota (Schüßler et al., 2001b), design of primers targeting protein-encoding genes, and covering the whole Glomeromycota phylum is challenging and may require the use of several primers (Stockinger et al., 2014). As discussed previously fungi belonging to the same family within the Glomeromycota can have a very contrasting nutrient exchange balance dependent on the plant colonized as well as the surrounding plants (Walder et al., 2012) reinforcing the observation that the symbiotic interaction between the plant and the fungus is highly dependent on the surrounding ecosystem (Gamper et al., 2010).

This symbiotic interaction is also dependent on the fungal community. For example, the balance of the exchange of nutrients between plant and fungus, with the plant providing photosynthetically fixed carbon in exchange for phosphate, varies from one pairing of plant-AMF to another, but is also dependent on the surrounding plants and fungi (Walder et al., 2012). The latter study indicated that the effect of the surrounding plants is possible

**BOX 7.1 Identification of Effective arbuscular mycorrhizal fungi (AMF) Assemblages**

To predict functional consequences in the host plant of different AMF, only by investigating the genome of the fungus, seems to be a tremendous task with the present state of knowledge and resources, given that we do not know what genes are making the differences in the plant. The effect depends not only on the assemblage of AMF isolates but also on the host plant, on the other soil microbes, to the mycelium network linking different plants and environmental conditions. However, it is known that the benefits the plant gains from AM symbiosis are not only the result of the exploitation of a bigger soil volume but also the impacts of changes in the plant defense mechanisms. These mechanisms are certainly anticipated by the priming of plant genes and therefore it might be possible to investigate the changes in proteomics and genomics of the host plant as the result of its infection by different AMF assemblages under the presence of a certain stress. These markers could be developed for the more important crops and major stresses in different regions of the world. These would make possible the development of molecular markers at the plant level, which could be used to identify, under field conditions, whether the assemblage of AMF colonizing a crop is appropriate for providing the best protection against the prevailing stress. This tool would be of major consequence in developing research into the effect of agricultural management practices on the functional consequences of their effects on the biodiversity of indigenous AMF populations.

through a common mycorrhizal network interconnecting two different plant species (Walder et al., 2012). The importance of an intact mycorrhizal network (Mommer et al., 2016) has also been demonstrated during plant sequence experiments in the presence of biotic or abiotic stresses (see Section 8.1). This shows that to obtain the correct assessment of the effect of the symbiotic interaction, the whole rhizosphere environment has to be taken into account, including the integrity of the mycorrhizal network, and not solely the biodiversity of the mycorrhizal fungi.

The diversity of Glomeromycota has been assessed in a large number of agronomic studies (Jansa et al., 2002; Gollotte et al., 2004; Galván et al., 2009; Brito et al., 2012b; Brígido et al., 2017) considering effects of biotic and abiotic factors on crops (Helgason et al., 1998; Johnson et al., 2004) over scales ranging from plot (Wolfe et al., 2006) to country and even larger areas (Öpik et al., 2009; Jansa et al., 2014; Bouffaud et al., 2015; Creamer et al., 2016). There have been several studies taking a more quantitative approach targeting ribosomal genes (Alkan et al., 2006; Pivato et al., 2007; Gamper et al., 2008; Jansa et al., 2008; Thonar et al., 2012; Schneider et al., 2015) or related biosynthesis activities (Vandenkoornhuyse et al., 2007). Farmer et al. (2007) reported that in one Chinese farming system mycorrhizal inoculation at planting increased crop yield as well as crop quality. In the case of sweet potato, an average crop yield increase of 17% was observed over a 2-year period. The quality of the crop improved, with a larger content of carotene and sugar, but this was very dependent on the specific mycorrhizal inoculum used. This example again suggests the importance of the mycorrhizal fungal diversity. Mycorrhiza have also been shown to improve tolerance to heavy metal stress (Rivera-Becerril et al., 2002; Evelin et al., 2009) but the link between diversity and functionality is still poorly understood.

The mycorrhizal symbiosis contributes to a multiplicity of ecosystemic services (Gianinazzi et al., 2010) and yet our limited knowledge of the functions involved are mainly based on experiments focused on a restricted number of model species of plant and fungus. Recent developments in the sequencing capacity for genome analysis open the possibility of analyzing these interactions in complex systems, such as the managed soil environment.

## 7.3 CONCLUSIONS

The effectiveness of the mycorrhizal symbiosis is very dependent on environmental conditions, but is also influenced by the community of higher plants as well as the local soil biota. At present, our ability to fully describe AMF diversity is still in a very early phase in terms of the taxonomic units present, despite the recent developments in sequencing capacity. Establishing the linkages between genetic and the functional diversity under field conditions with biotic and abiotic stress is more challenging. If we are to capitalize on the potential benefits from managing AMF diversity, there needs to be the capability of designing solutions that are appropriate in different agroecosystems and can be applied on a site-by-site basis. Identifying the molecular markers that will allow quantitative assessments to be made of the potential to exploit indigenous AMF is crucial.

Short-term considerations suggest that identifying key markers in major crops for functionally effective AMF assemblages against major biotic and abiotic stresses should be given more prominence.

Chapter 8

# Management of Biological and Functional Diversity in Arbuscular Mycorrhizal Fungi Within Cropping Systems

## Chapter Outline



In the preceding chapters we have outlined the benefits that host plants can accrue from forming arbuscular mycorrhiza (AM). These advantages are well established in the literature and range from nutrient acquisition by the plants, the protection against abiotic stresses, such as drought, salinity, toxicity of metals and metalloid ions, and enhanced defense against biotic stress, such as the invasion of root pathogens. Moreover, AM play another key role in providing ecosystem services through improved soil quality and soil protection through their effects on soil aggregation and associated physical properties. There is also evidence of other benefits from arbuscular mycorrhizal fungi (AMF) to society that include protection of the quality of water resources, by reducing $NO_3^-$ and $NH_4^+$ leaching, and climate, by reducing the emission of $N_2O$ gas (see Section 3.1.1). All these aspects are crucial for the sustainable intensification of agriculture, both in countries from different parts of the world, where injudicious use of manure and excessive application of fertilizers and pesticides are imposing unacceptable environmental impacts, and in developing regions, where there is an urgent need to improve land productivity but which suffer from a scarcity of resources (see Chapter 1). However, the intentional use of AMF within agricultural


**Functional Diversity of Mycorrhiza and Sustainable Agriculture.**
**DOI: http://dx.doi.org/10.1016/B978-0-12-804244-1.00008-3**

cropping systems has received little attention, especially for large-scale field crops. There are three key aspects that prevent the full exploitation of AMF in agricultural ecosystems:

1. For each of the benefits from AM there seems likely to be functional diversity between different AM fungal species and even between isolates of the same taxonomic unit.
2. Time is required to achieve an effective colonization, which is likely to be crucial when bioprotection is a major benefit of AMF (Brito et al., 2014) but it also seems important in relation to nutrient acquisition (Miller, 2000).
3. The high cost of commercial inoculum coupled with its lack of biological diversity, given that it is usually developed from a few isolates of species that readily produce an abundance of spores. But more importantly, the evidence suggests that introduced AMF may not provide the benefits that could be obtained if the indigenous population were to be exploited (see Sections 3.2, Box 3.1 and Box 5.1).

These concerns clearly establish that there is a need to develop strategies for the management of indigenous AMF within different cropping systems, specifically focusing on how to manage the AMF biological diversity in the roots of the crops and what can be done to enhance early colonization. A number of studies have identified concerns and opportunities to enhance these aspects of cropping systems (Table 8.1). All indicate that reduction in the intensification of the farming system is of paramount importance. This includes the reduction of inputs in general. No-till and reduced tillage, conservation tillage systems are all seen as contributing to greater AMF presence and functioning. Monocropping and the frequent repetition of cereals and nonmycorrhizal crops in a rotation are considered to be deleterious for AMF. Bare fallow and short fallows are considered to have a negative impact, as is the application of manure slurry. Solid or farmyard manure is considered to be beneficial as is the return of crop residues. The inclusion of legumes in a rotation or as cover crops is also identified as a positive approach to supporting AMF.

When intact extraradical mycelium (ERM) acts as the preferential propagule, the colonization of the plant roots starts earlier and develops faster than from any other source of propagules (see Section 6.2). The ERM can be developed in the soil by different elements of the cropping system, such as previous crops, cover crops, or weeds and kept intact at the seeding of the following crop if appropriate tillage techniques, e.g., no-till or reduced shallow cultivation, are adopted. There is plenty of evidence that there is some degree of specificity between plants and mycorrhizal species and different plant functional groups, such as forbs and grasses harbor different AMF populations (Torrecillas et al., 2012; Lekberg et al., 2013). Scheublin et al. (2004) found that AMF communities vary between plant functional groups (legumes and nonlegumes) and between plant species. It is also notable that

**TABLE 8.1** Design of cropping systems to enhance arbuscular mycorrhizal fungi (AMF) diversity and deliver functional benefits to crops and ecosystem services

| Crops That Support | Crops That Minimize | Supportive Practices | Cropping Systems | Other Negative Practices | References |
|---|---|---|---|---|---|
| Grain legumes, maize, sunflower, potato | Cereals, especially if frequent in rotation,<br>Brassicacea, such as Canola, mustard cover crops<br>Chenopodiacea, such as sugar beet and fodder beet | No-till<br>Reduced intensity tillage systems<br>Reduced inputs of fertilizer, pesticides, growth regulators,<br>Constant ground cover with crops and residues | Multiple cropping<br>Intercropping<br>Alley cropping<br>Cover cropping<br>Avoid monocropping | Clean fallow<br>Short fallow | Plenchette et al. (2005) |
| Oat | | Incorporation of oat straw | Avoid residual herbicides | | Alguacil et al. (2010) |
| Legume cover crop | Monoculture | Residue returns, no external inputs (fertilizers, pesticides, or organic amendments) | Increased number of crops in rotation | | Tiemann et al. (2015) |
| Mycotrophic cover crops | Nonmycotrophic crops and cover crops | No-till and ridge-till | Cover cropping | Fallow | Kabir (2005) |
| | | Reduced tillage<br>Extended crop rotations | Organic systems with reduced tillage<br>Avoid organic systems with conventional tillage | Slurry application | Säle et al. (2015) |
| Cover crops; legumes; C3, C4 grasses; native plants | Brassicacea, such as mustard, *Brassica rapus* cover crops | Reduced or no-till for cover crops | Perennial agriculture | Herbicides and fungicides | Vukicevich et al. (2015) |

Ellouze et al. (2013) even found AMF biological and functional diversity associated with different chickpea cultivars. Vierheilig (2004) found that further root colonization by AMF in already mycorrhizal plants is suppressed after a critical level of root colonization. The evidence from numerous authors has shown that in a plant succession, when the ERM associated with the first plant species is kept intact it promotes a faster and earlier colonization of the subsequent crop. It is therefore plausible that in these circumstances the AMF diversity associated with a first plant will influence the AMF diversity present in the roots of the second plant. In Section 4.3.2, we showed how important the ability to colonize across plant functional groups can be in allowing the use of Developer plants to establish more effective assemblages of AMF, in that case for protection against Mn toxicity, and importantly to develop an ERM network in the soil ahead of planting the next, sensitive crop. Therefore, the strategy we are presenting, which aims to overcome limitations in the intentional use of AMF within the cropping systems and to improve the benefits that result, is to manage the indigenous AMF through intermediary mycotrophic plant species. Ideally we need to grow as intermediary hosts the plants that are well adapted to the prevailing stresses in the ecosystem and are able to support and select for the AMF assemblage developing ERM, which, if kept intact by appropriate tillage, would provide protection of the cash crop at the time of planting (see Section 4.3.2). Thus it might be possible not only to enhance the colonization of the second crop but also to influence the AMF diversity colonizing its roots. In this chapter we will consider some case studies and discuss their possible application in different cropping systems in different regions of the world.

## 8.1 MANAGING INDIGENOUS AMF IN AGROECOSYSTEMS

The contributions that AM can make to enhancing yields but reducing inputs are likely to become most important in relation to protection of crops from abiotic and biotic stresses. Most of the examples selected for discussion come from detailed experiments at the greenhouse scale.

### 8.1.1 Managing Indigenous AMF to Overcome Abiotic Stresses

Although the ability of AM to protect host plants from abiotic stresses is well documented, approaches to the management of indigenous AMF for this purpose have not received the same attention. We explore two aspects; the protection against toxic levels of manganese and the improved exploitation of nutrient resources.

> *When an excessive level of metalloid ions is the main stress on host plants, AMF colonization starting preferentially from intact ERM is crucial to the level of bioprotection granted.*

Protection of susceptible mycotrophic plants against manganese toxicity is best illustrated in experiments reported by Brito et al. (2014) and Alho et al. (2015) (see Sections 4.3.2 and 5.1.1). Wheat (*Triticum aestivum* L., var. Ardila) and subterranean clover (*Trifolium subterraneum* L. cv Nungarin), respectively, were the susceptible host plants investigated in these two studies. Although the research was carried out in pot experiments under controlled environments, it identifies key requirements for successful field application. In both studies the pot experiment was carried out in two phases. The field soil used was unsterilized and contained excessive levels of manganese that impaired growth of both wheat and subterranean clover. Two factors were studied: ERM developer species, grown in the first phase of the experiment, and the integrity of the ERM present at the beginning of the second phase. In Phase 1 of the experiment, four plant species, which occurred naturally in the ecosystem developed on the soil used in the two experiments, were selected and grown, two (*Silene gallica* L., *Rumex bucephalophorus* L.) being non or scarcely mycotrophic and two (*Lolium rigidum* G. and *Ornithopus compressus* L.) were mycotrophic. These four plants were designated as "Developers" of ERM. At the end of Phase 1, all the Developer plants were killed with a systemic herbicide to ensure that the means of ending this phase was not a factor in the experiment. In Phase 2 of each experiment, the level of integrity of the ERM was varied by mechanical disturbance of the soil. In half of the pots the soil was passed through a sieve (the Disturbed treatment), which fragmented the ERM developed by any mycorrhiza associated with the Developers. In the remaining pots there was no mechanical disturbance (the Undisturbed treatment), which kept intact any ERM formed by mycorrhiza on the Developers. Wheat or subterranean clover was then planted in all pots. Therefore, at the time when both crops were growing, indigenous AMF were always present but colonization started from different types of propagule. In the Disturbed and Undisturbed treatments following the growth of nonmycotrophic developers, the predominant propagules were spores because no mycorrhizal development was expected on nonmycotrophic plants. In the Disturbed soil treatments after mycotrophic developers, the propagules could be disrupted ERM together with spores, and colonized root fragments. Only in the Undisturbed soil treatment, following the growth of mycotrophic developers, were intact ERM, spores, and colonized root fragments all present as propagules.

The aim of both experiments was to develop a strategy for improving the efficacy of indigenous AMF in protection of cultivated plants against an abiotic stress present at planting. The specific objectives were (1) to evaluate the significance of the source of propagules for AMF (spores, colonized root fragments, or intact ERM) and (2) evaluate the possibility of managing AMF biological diversity associated with the roots of the susceptible cultivated plant by choice of Developer plant species. It was expected that the AMF diversity present in the roots of the second plant in the succession, in this case the susceptible host, would be influenced by that of the first plant

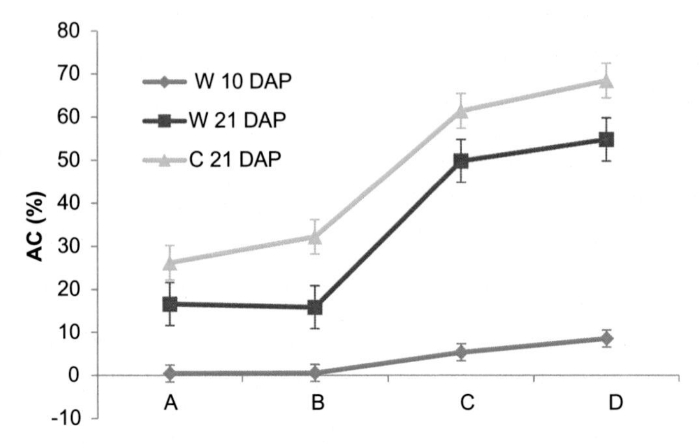


**FIGURE 8.1** Arbuscular colonization (AC) of wheat by indigenous AMF 10 days (W 10 DAP), 21 days (W 21 DAP), and subterranean clover 21 days after planting (C 21 DAP). Colonization was initiated by different propagules. A, Spores only; B, Spores, fragmented mycelium and colonized root fragments of *Lolium rigidum* or *Ornithopus compressus*; C, Spores and intact extraradical mycelium (ERM) associated with the roots of *L. rigidum*; and D, Spores and intact ERM associated with the roots of *O. compressus*. The soil contained 22.6 mg Mn $kg^{-1}$.

species of the succession, the Developer, if intact ERM was the preferential propagule initiating AMF colonization.

The AMF colonization started earlier and developed faster when an intact ERM was the preferential source on inoculum (Fig. 8.1, line W10 DAP, propagules C and D) and significantly improved the growth of both plants in the presence of excessive levels of Mn in the soil (Fig. 8.2 and Plates 8.1 and 8.2). It is notable that no differences in shoot growth were found between the treatment with AMF colonization resulting from spores alone and that from spores together with colonized root fragments. It is commonly identified in the literature that, when the role of AMF is bioprotection of plants against abiotic or biotic stresses mycorrhiza must be formed and be well-established before it comes into contact with the stressor if a high level of protection is to occur (Rufyikiri et al., 2000; Petit and Gubler, 2006; Nogales et al., 2009). In these two experiments, the stress (toxic concentration of Mn) was present in the soil at the time of planting, which is the typical situation under field conditions. Consequently the initial rate of AMF colonization would be critical. Apparently, this could only result and an adequate level of protection be achieved, when an intact ERM was the preferential propagule, specifically if the two crops were planted after either *O. compressus* or *L. rigidum* in the Undisturbed soil treatment. Only in these two treatments was there a significant reduction in the concentration of manganese in the shoots of wheat and roots of subterranean clover achieved (Fig. 8.3). The Mn concentration in the subterranean clover shoots was smaller than that in wheat, and was not significantly affected by the treatments, which is consistent with

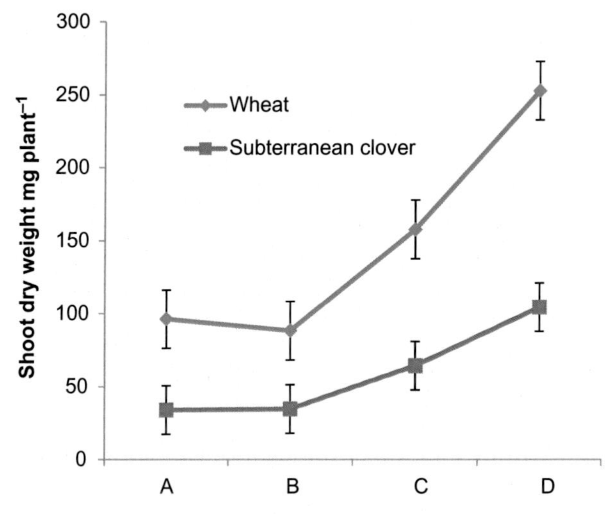


**FIGURE 8.2** Shoot dry weight of wheat (blue line) and subterranean clover (red line) 21 days after planting. Colonization by indigenous AMF was initiated by different propagules. A, Spores only; B, Spores, fragmented mycelium, and colonized root fragments of *Lolium rigidum* or *Ornithopus compressus*; C, Spores and intact extraradical mycelium (ERM) associated with the roots of *L. rigidum*; and D, Spores and intact ERM associated with the roots of *O. compressus*. The soil contained 22.6 mg Mn $kg^{-1}$.

**PLATE 8.1** Growth of wheat in soil containing excessive levels of Mn following colonization initiated by different propagules. A, B, and D are the propagule treatments indicated in Fig. 8.1 (A, Only spores; B, Spores, fragmented mycelium, and colonized root fragments of *Ornithopus compressus*; and D, Spores and intact ERM of the indigenous AMF associated with *O. compressus*). The protection of wheat plants by arbuscular mycorrhizal fungi (AMF) was only conferred when an intact extraradical mycelium (ERM) was present in the soil at the time of planting.

**PLATE 8.2** Growth of wheat in soil containing excessive levels of Mn following colonization initiated by similar types of propagule from indigenous arbuscular mycorrhizal fungi (AMF). C and D are the propagule treatments shown in Figs. 8.1, 8.4 (C, Spores and intact extraradical mycelium (ERM) of the indigenous AMF associated with *Lolium rigidum*; D, Spores and intact ERM of the indigenous AMF associated with *Ornithopus compressus*). The functional consequences of the different AMF associated with the two Developer plants are evident.

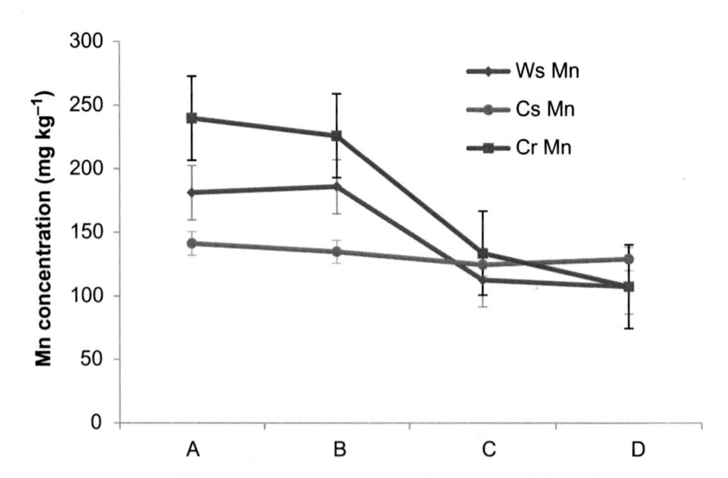


**FIGURE 8.3** Effect of the propagule type from indigenous AMF in the soil at the time of planting on manganese concentration in wheat shoots (Ws Mn) and subterranean clover shoots (Cs Mn) and roots (Cr Mn), 21 days after planting. Colonization by indigenous AMF was initiated by different propagules: A, Spores only; B, Spores, fragmented mycelium, and colonized root fragments of *Lolium rigidum* or *Ornithopus compressus*; C, Spores and intact extraradical mycelium (ERM) associated with the roots of *L. rigidum*; and D, spores and intact ERM associated with the roots of *O. compressus*. The soil contained 22.6 mg Mn $kg^{-1}$.

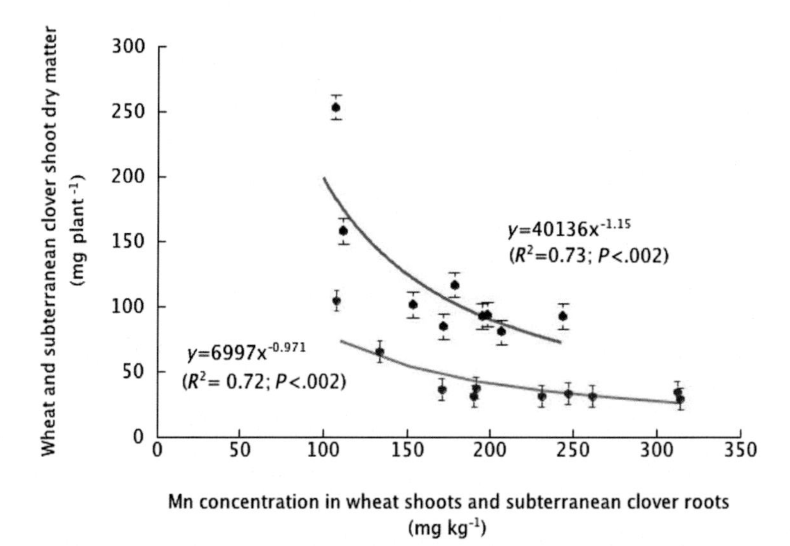


**FIGURE 8.4** Relationship between shoot dry weight, 21 days after planting and Mn concentration in the shoots of wheat (blue line and markers) or roots of subterranean clover (red line and markers). The soil contained 22.6 mg Mn $kg^{-1}$.

subterranean clover being relatively tolerant to Mn when it is not depending on biological nitrogen fixation.

Although there were no significant differences between the two mycotrophic Developer plant species on the colonization of the following crop (Fig. 8.1) or the concentration of Mn (Fig. 8.3), the growth of wheat (Fig. 8.2 and Plate 8.2), and subterranean clover (Figs. 8.2 and 8.5(3)) were significantly enhanced after *O. compressus* relative to that observed after *L. rigidum*. The growth of wheat and subterranean clover was related significantly but negatively to the concentration of Mn in the shoots and roots, respectively (Fig. 8.4). In the case of subterranean clover the protection conferred, when an intact ERM was the main source of inoculum, was primarily to the bacteroids in the root nodules (Fig. 5.7 and Fig. 8.5(1)). This led to greater N acquisition (Fig. 5.7 and Fig. 8.5(2)) as the result of enhanced biological nitrogen fixation. This observation is consistent with the subterranean clover being more sensitive to excess levels of metalloid ions when N is provided by biological fixation (see Section 5.1.1). In addition to the protection against excessive levels of manganese, the presence of an intact ERM at the time of planting enhanced the uptake of important nutrients for plant growth, such as P and S (Fig. 8.6).

*In a succession of plants it appears possible to manage the AMF biological diversity present in the roots of the second plant if intact ERM is the preferential initiator of colonization.*

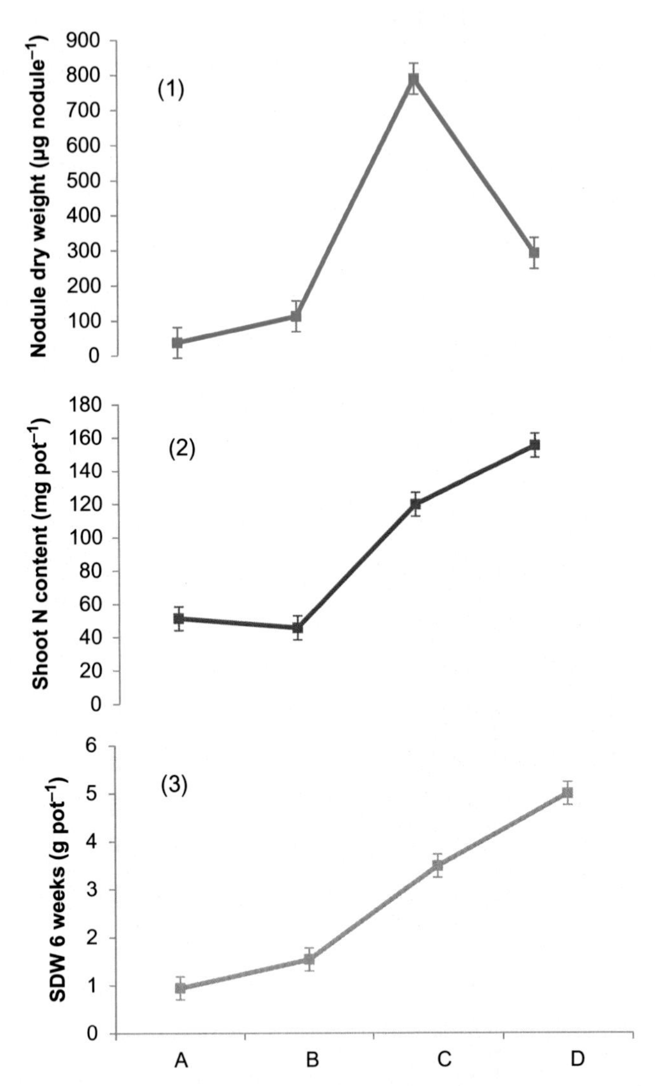


**FIGURE 8.5** Effect of the indigenous arbuscular mycorrhiza (AM) on (1) nodule dry weight, (2) shoot N content, and (3) shoot dry weight (SDW) of subterranean clover 6 weeks after planting. Colonization by indigenous AMF was initiated by different propagules: A, Spores only; B, Spores, fragmented mycelium and colonized root fragments of *Lolium rigidum* or *Ornithopus compressus*; C, Spores and intact extraradical mycelium (ERM) associated with the roots of *L. rigidum*; D, Spores and intact ERM associated with the roots of *O. compressus*. The soil contained 22.6 mg Mn $kg^{-1}$.

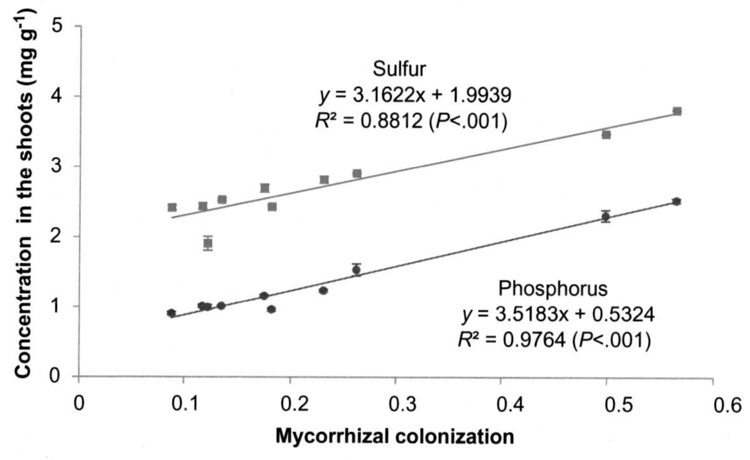


**FIGURE 8.6** Relationship between mycorrhizal colonization and P (red line) and S (blue line) concentration in the shoots of wheat at 21 days after planting. The soil contained 22.6 mg Mn $kg^{-1}$.

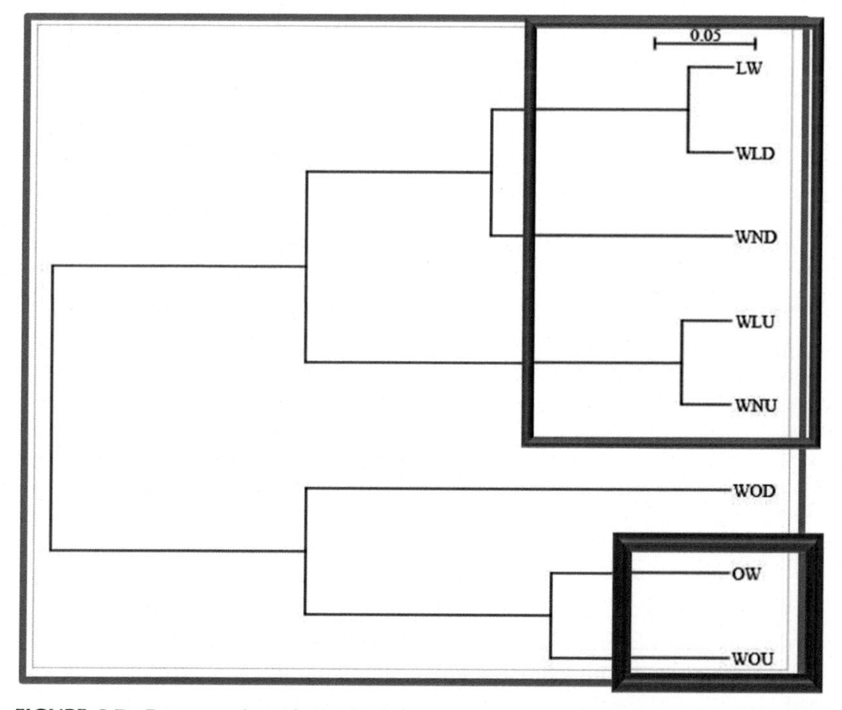


**FIGURE 8.7** Representation of the similarities between the community structures of AMF present in *wheat* roots in succession to *Ornithopus compressus* or *Lolium rigidum* plants with or without soil disturbance, evaluated by 454-pyrosequencing technique. OW, *O. compressus*, only spores as AMF propagules; LW, *L. rigidum*, only spores as AMF propagules; WLD, wheat after *L. rigidum*, Disturbed soil; WLU, wheat after *L. rigidum*, Undisturbed soil; WNU, wheat, only spores as AMF propagules; WOD, wheat after *O. compressus*, Disturbed soil; WOU, wheat after *O. compressus*, Undisturbed soil (after Brígido et al., 2017).

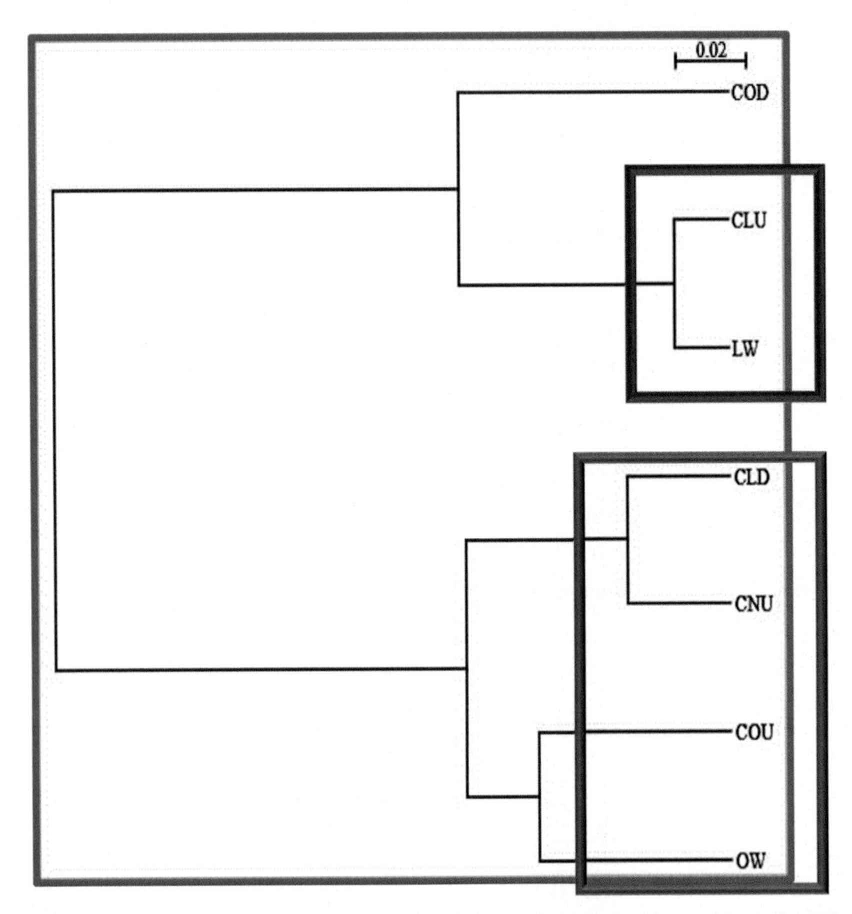


**FIGURE 8.8** Representation of the similarities between the community structures of AMF present in roots of subterranean clover in succession to *Ornithopus compressus* or *Lolium rigidum* plants with or without soil disturbance, evaluated by 454-pyrosequencing technique. CLD, subterranean clover after *L. rigidum*, Disturbed soil; CLU, subterranean clover after *L. rigidum*, Undisturbed soil; CNU, subterranean clover, only spores as AMF propagules; COD, subterranean clover after *O. compressus*, Disturbed soil; COU, subterranean clover after *O. compressus*, Undisturbed soil; OW, *O. compressus*, only spores as AMF propagules; LW, *L. rigidum*, only spores as AMF propagules (after Brígido et al., 2017).

A notable aspect of these studies was the considerably increased growth of wheat (Plate 8.2; Fig. 8.2) and subterranean clover (Figs. 8.2 and 8.5(3)) in the Undisturbed soil treatment (ERM intact) when the ERM was developed on *O. comp*ressus in relation to that developed by *L. rigidum*. Under the condition with an intact ERM, the AMF diversity present in the roots of wheat (Fig. 8.7) and subterranean clover (Fig. 8.8) was affected by the previous plant. When spores, fragmented mycelium and colonized root fragments were the main AMF propagule source, *O. compressus* (Treatment OW) and *T. subterraneum* (Treatment CNU) (both belonging to the *Fabaceae*) or

*L. rigidum* (Treatment LW) and *T. aestivum* (Treatment WNU) (both belonging to the *Poaceae*) present similar communities of AMF within each plant species group and clear differences could be detected between the two botanical groups, indicating preferential symbioses between the AMF population and the plant species group, at least at the family level. However, the presence of an intact ERM associated with the first plant (LW or LC and OW or OC), acting as the preferential source of AMF propagule, shifts the AMF community present in the second plant (*T. subterraneum* or *T. aestivum*) corresponding to the one in the first plant of the succession independently of their botanical group. Therefore, it is reasonable to assume that the better growth of both wheat and subterranean clover after *O. compressus* in the Undisturbed soil treatment (intact ERM) are probably related to functional consequences of the AMF diversity associated with the roots of each of these two crops imposed by the Developer plant.

We conclude that to improve the role of AMF in the bioprotection of cultivated plants against abiotic stress, a strategy for their management within agricultural systems requires that:

1. ERM needs to be developed previously in the soil in conjunction with mycotrophic plant species that are tolerant to the stress. This can be achieved using previous crops in the crop rotation, cover crops or even weeds. The best strategy would take account of the cropping system and environmental conditions.
2. ERM must be intact in the soil when the crop to be protected is seeded. This will also need the adoption of appropriate tillage techniques, such as no-till or superficial reduced tillage systems.
3. AMF diversity present in the roots of the crop to be protected needs to be managed by the right choice of the Developer plants to take advantages of possible functional consequences.

*AMF colonization starting from intact ERM enhances the role of AM symbiosis in nutrient acquisition by the host plant.*

*Managing indigenous AMF to improve nutrient acquisition* came to prominence in the development of an understanding of the early P nutrition of maize grown under no-till by Miller and coworkers at the University of Guelph, Canada (Miller, 2000). The research established the importance of the presence in the soil of a previously developed and intact ERM for early absorption of P (Kabir et al., 1998; Miller, 2000), even under conditions when P fertilizer was being applied and the concentration of P in the soil was greater than that for which a plant growth response would be expected. It was also applicable even when the presence of an intact ERM apparently did not affect the level of colonization in maize (Miller et al., 1995a). Here we consider the work of Brito et al. (2013b, 2014), with wheat as the host crop, together with unpublished results for maize. These authors carried out a two-phase pot experiment. In Phase 1 a mixture of naturally occurring grass weeds, *L. rigidum*, *Avena sterilis* L., and *Phalaris minor* Retz., commonly associated with wheat fields in the Alentejo Region of Portugal, was grown in sieved, unsterilized field soil for 4 weeks. The weeds were then controlled, either by an application of systemic herbicide to keep intact the ERM of AMF associated with the weed roots, or by soil disturbance (ERM disrupted). In Phase 2, wheat was planted (4 plants pot$^{-1}$) and allowed to grow for 3 weeks. One week after planting the wheat, nitrogen (50 mg N kg$^{-1}$ dry soil), zinc (3.4 mg Zn kg$^{-1}$ of dry soil), and sulfur (1.7 mg kg$^{-1}$ of dry soil) were added to the pots. Wheat plants were harvested 21 days after planting. In the experiment with maize the same approach was used, but *O. compressus* was grown for 8 weeks during Phase 1 of the experiment to allow an extensive ERM to develop in the soil. Bicarbonate extractable P in the soil at the beginning of the experiment was 7 mg kg$^{-1}$ of soil. This time shoots of *O. compressus* were then removed by cutting below the soil surface. Two levels of ERM colonization potential were created: ERM intact (Undisturbed soil treatment) and ERM disrupted (Disturbed soil treatment). For Phase 2, maize was planted, 2 plants pot$^{-1}$, and 4 levels of phosphorus were applied (0, 6, 12, and 18 mg P kg$^{-1}$ dry soil equivalent to 0, 15, 30, and 45 kg P ha$^{-1}$). All pots received the same amount of calcium (1000 mg kg$^{-1}$ dry soil as dolomitic limestone with 10% of Mg equivalent to 2500 kg ha$^{-1}$), nitrogen (ammonium nitrate) (63 mg N kg$^{-1}$ dry soil ~158 kg N ha$^{-1}$), potassium (potassium sulfate) (31 mg K kg$^{-1}$ dry soil, ~78 kg K ha$^{-1}$), and zinc (zinc sulfate) (3.8 mg Zn kg$^{-1}$ dry soil ~10 kg Zn ha$^{-1}$). Maize plants were harvested 21 and 31 days after planting.

The presence of an intact ERM associated with roots of grass weeds at the time of planting increased the rate of AMF colonization, biomass, and the P content of shoots in wheat, after 3 weeks growth (Fig. 8.9). As the initial value of available P in the soil was very small (7.9 mg kg$^{-1}$ soil) and the weeds grew vigorously during Phase 1 of the experiment (2.8 g of shoot dry matter per pot) availability of nutrients, including P, in the soil was reduced. The ability of indigenous AMF to improve P absorption by the wheat under semiarid Mediterranean environment was also found by Smith et al. (2015) both in pot and field experiments. Although the arbuscular colonization was relatively small (7%) compared to that shown in Fig 8.9 when an intact ERM was present at the time of wheat planting (~30%) the shoot P absorption via AMF pathway was ~12% and these authors concluded that "AM fungi should not be ignored as contributors to wheat P uptake in the field and that the symbiosis should be taken into account in any search to improve the P uptake efficiency of wheat and other cereals."

We can conclude, therefore, that:

1. Under conditions where availability of P is small, naturally occurring mycotrophic weeds can be used to develop ERM of indigenous AMF in the soil.

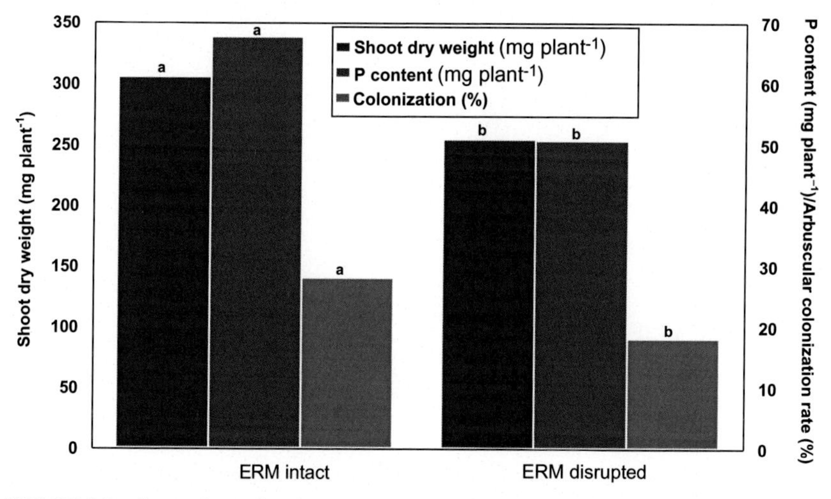


**FIGURE 8.9** Comparison of an intact extraradical mycelium (ERM) with ERM disrupted on, shoot dry weight (blue bars), P content (mauve bars), and arbuscular colonization rate (red bars), in wheat. Results show the average for 21 days after planting. For each parameter the same letter indicates values are not significantly different at $P = .05$ (after Brito et al., 2013b).

2. Adoption of appropriate tillage technique, which maintains the integrity of the ERM, can be used to improve the contribution of AMF to nutrient acquisition of cultivated plants within a cropping system.

The content of major nutrients in wheat plants also increased (Fig. 8.10), when grown in soils with excessive levels of manganese, but only in the presence of an ERM that had developed previously in association with mycotrophic tolerant plants and kept intact at the planting of the susceptible crop. The benefit to the susceptible wheat crop is therefore not only the alleviation of manganese toxicity that typically impaired the growth of crops but also the improvement in the ability to access a greater nutrient supply. Importantly, soil acidity constrains crop production on about 70% of the potential arable land across the world (Haug, 1984), mainly due to the associated toxicity of Al and Mn to plant growth (Mora et al., 2004). Furthermore many of those areas are in tropical regions of the world where access to inputs, such as limestone and fertilizer, is greatly limited. The strategy of developing ERM from indigenous AMF by mycotrophic tolerant plants and keeping it intact at seeding of susceptible crops, is not only a strategy to alleviate the toxicity of the metalloid ions, but can also enhance access to and acquisition of important scarce nutrients by crops.

In the experiment with maize, when an intact ERM was the preferential AMF propagule in the soil, arbuscular colonization (AC) was greatly enhanced and, even at the largest P application rate, the value was above 60% (Fig. 8.11). When AC was the parameter used to evaluate the progress of AMF colonization, P application to the soil appeared to influence the rate

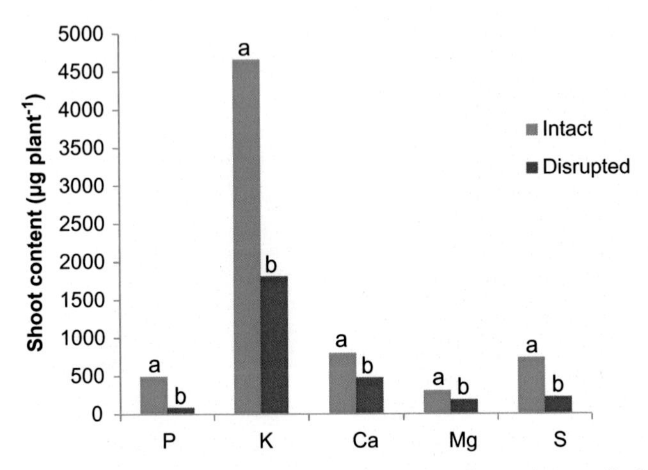


**FIGURE 8.10** Comparison of the presence of an intact with a disrupted extraradical mycelium (ERM) at planting on the content of alkaline elements and S in wheat. Results show the values for 21 days after planting. For each parameter the same letter indicates values are not significantly different at $P = .05$. The condition of the ERM was created by previously growing mycotrophic Developer plants (*Ornithopus compressus* L. or *Lolium rigidum* G.) and applying differential soil disturbance treatments (Undisturbed soil, ERM kept intact; Disturbed soil, ERM disrupted) (after Brito et al., 2014). The soil contained 22.6 mg Mn $kg^{-1}$.

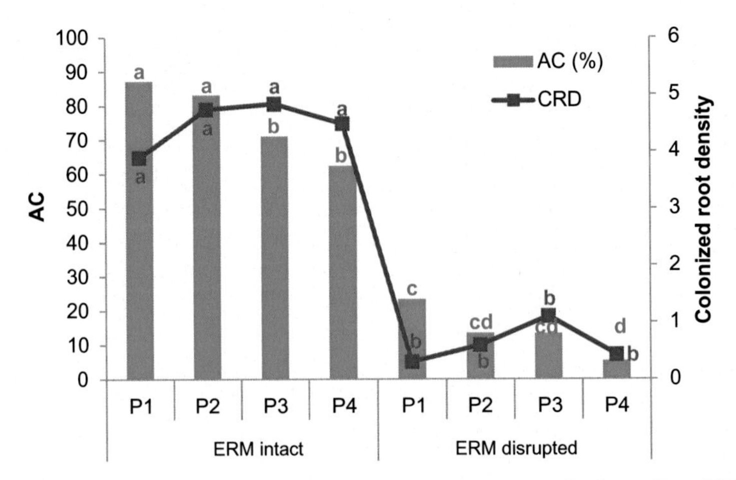


**FIGURE 8.11** Effect of the presence of an intact or disrupted extraradical mycelium (ERM) at the time of maize planting on AMF arbuscular colonization % (AC, blue bars) and colonized root density (CRD cm of colonized root $cm^{-3}$ of soil, red squares), under four P levels applied to the soil (P1, P2, P3, and P4 stands for 0, 6, 12, and 18 mg P $kg^{-1}$ of dry soil, respectively). ERM was previously developed in the soil by *O. compressus* and indigenous AMF. Results show the means of values for 21 and 31 days after planting. For each parameter the same letter indicates values are not significantly different at $P = .05$.

negatively. However, when colonization was evaluated by colonized root density (CRD) (see Box 2.2), the level of P applied had no significant effect on AMF colonization of the maize root system, whatever propagules were in the soil at planting. Maize benefitted from the presence of an intact ERM in the soil developed previously in association with *O. compressus*. Shoot dry weight and P content (Fig. 8.12) increased significantly at each P level in the Undisturbed soil. Although the plant was growing faster in Undisturbed soil, the concentration in the shoot of N, K (Fig. 8.13), Ca, and S (Fig. 8.14) remained the same but uptake of these nutrients was significantly improved. The beneficial effect of an intact ERM on the growth of the maize roots was also evident (Plate 8.3). All these results are consistent and confirm that the role of indigenous AMF can be greatly enhanced if an intact ERM is the preferential AMF propagule. Under these varied conditions plant response to P fertilizer and uptake of other important nutrients can all be improved by the earlier colonization stemming from the presence of an intact ERM at the time of seeding.

> *Bioprotection of host plants against soil fungal pathogens seems to require AMF colonization being started by intact ERM.*

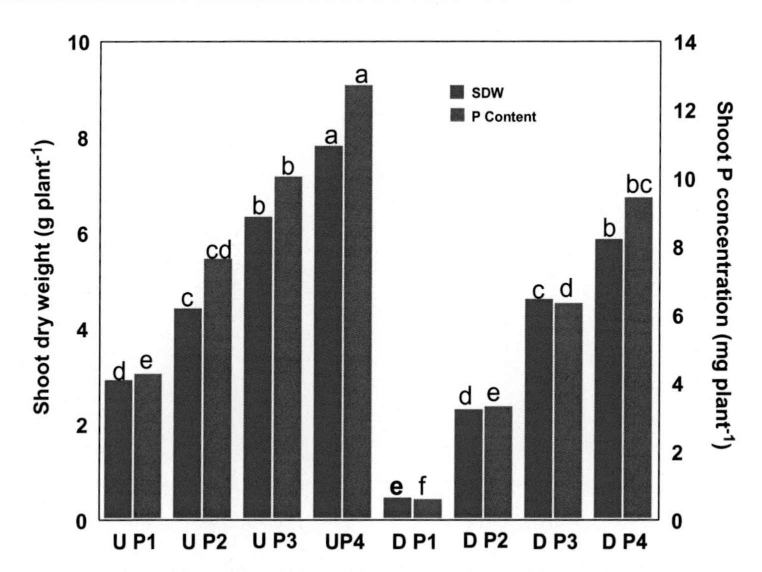


**FIGURE 8.12** Effect of the presence of intact or disrupted extraradical mycelium (ERM) of arbuscular mycorrhizal fungi (AMF) at the time of maize planting on the shoot dry weight (SDW) and P content, under four P levels applied to the soil (P1, P2, P3, and P4 stands for 0, 6, 12, and 18 mg P $kg^{-1}$ of dry soil, respectively). U (Undisturbed soil) and D (Disturbed soil) corresponds to ERM being intact or disrupted, respectively. Results show the means of values for 21 and 31 days after planting. ERM was previously developed in the soil by *Ornithopus compressus* and indigenous AMF. For each parameter the same letter indicates values are not significantly different at $P = .05$.

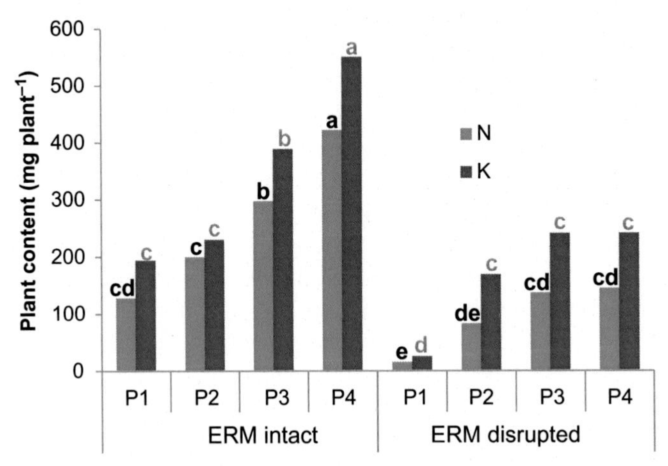


**FIGURE 8.13** Effect of the presence of an intact or disrupted extraradical mycelium (ERM) at the time of maize planting on the shoot N and K content, under four P levels applied to the soil (P1, P2, P3, and P4 stands for 0, 6, 12, and 18 mg P $kg^{-1}$ of dry soil, respectively). Results show the means of values for 21 and 31 days after planting. ERM was previously developed in the soil by *Ornithopus compressus* and indigenous arbuscular mycorrhizal fungi (AMF). For each parameter the same letter indicates values are not significantly different at $P = .05$.

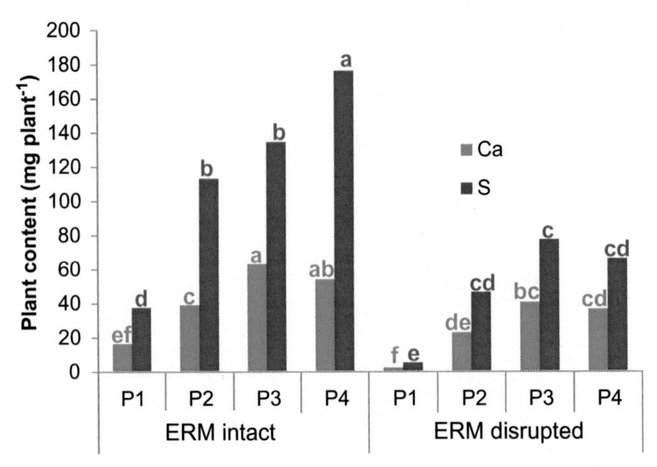


**FIGURE 8.14** Effect of the presence of intact or disrupted extraradical mycelium (ERM) at the time of maize planting on the shoot content of Ca (blue bars) and S (red bars), under four P levels applied to the soil (P1, P2, P3, and P4 stands for 0, 6, 12, and 18 mg P $kg^{-1}$ of dry soil, respectively). Results show the means of values for 21 and 31 days after planting. ERM was previously developed in the soil by *Ornithopus compressus* and indigenous arbuscular mycorrhizal fungi (AMF). For each parameter the same letter indicates values are not significantly different at $P = .05$.

**PLATE 8.3** Effect of the presence of an intact extraradical mycelium (ERM) of native arbuscular mycorrhizal fungi (AMF) developed in association with *Ornithopus compressus* on the growth of the roots of maize 21 days after planting (related to data presented in Fig. 8.11). (U P1, Undisturbed soil treatment; ERM intact and no P applied to the soil; D P1, Disturbed soil treatment; ERM disrupted and no P applied to the soil).

### 8.1.2 Managing Indigenous AMF to Overcome Biotic Stresses

In most agricultural ecosystems, soilborne plant pathogens can be a major limitation to production. Although losses are difficult to estimate, in the United States it may be that pathogens are responsible for as much as 50% of total crop loss (Lewis and Papavizas, 1991). The need for alternative strategies to reduce the impact of soilborne pathogens on crop productivity is increasing for economic reasons and as a means of addressing environmental concerns over the use of chemical treatments. AMF are known to be effective in reducing the infection and damage to the plant cause by several soilborne fungal diseases, including *Fusarium oxysporum*. However, common approaches have required AMF to be inoculated in advance of exposure to the disease organisms. For example, to control onion white rot (*Sclerotium cepivorum* Berk.) Torres-Barragán et al. (1996) showed that if the AMF isolate, *Glomus* sp. Zac-19, and *S. cepivorum* were inoculated at the same time, there was no effect on disease incidence (DI), but positive effects were found when the onions were inoculated with this AMF at the nursery stage, before the plants contact with the disease organism. However, this approach is not possible for crops that are seeded into the field. Furthermore, even for transplanted crops, it requires the use of cultured AMF inoculum, which is likely less diverse and not adapted to other local conditions, rather than indigenous AMF (see Box 3.1).

Protection against the soilborne disease *Fusarium oxysporum* f.sp. *radicis-lycopersici* in tomato plants *(Lycopersicon esculentum)* provides an example of an alternative, field-based approach to overcome a biotic stress. In this study again using a two-phase pot experiment under controlled conditions, Phase 1 involved the establishment of a Developer plant in unsterilized soil to create an ERM formed by indigenous AMF. In this case, as the Developer plant needed to be mycotrophic and nonsusceptible to the disease organism, *L. rigidum* was selected as the ERM Developer and grown for 8 weeks. At the end of this period it was treated with a systemic herbicide. Half of the pots were then either sieved (Disturbed soil treatment) to disrupt the ERM and in the remaining pots the soil was untouched (Undisturbed soil treatment) so that the ERM remained intact. Three pregerminated tomato seedlings were planted per pot and *F. oxysporum* f.sp. *radicis-lycopersici* was inoculated to the roots of each plant at planting, using 1 mL plant$^{-1}$ of a suspension with a concentration of 0, $10^3$, $10^6$, or $10^9$ conidia mL$^{-1}$. Nutrients were added to the soil at the planting of tomato to ensure that the strategy was compatible with the high level of fertilizer application, i.e., the current agriculture practice for tomato production. All pots received the same amount of Ca (1000 mg kg$^{-1}$ dry soil as dolomitic limestone with 10% Mg, equivalent to 2600 kg Mg ha$^{-1}$), N (7.9 mg N kg$^{-1}$ as ammonium nitrate, ~20 kg N ha$^{-1}$), K (32 mg K kg$^{-1}$ as potassium sulfate, ~80 kg K ha$^{-1}$), Zn (3.8 mg Zn kg$^{-1}$ as zinc sulfate, ~10 kg Zn ha$^{-1}$), and B (3.8 μg kg$^{-1}$ as sodium borate). Tomato was grown for 3 weeks before shoot dry weight, AC and DI were evaluated. DI was assessed visually (see Plate 8.4), being graded from no visible symptoms (Scored 1) to stem fully affected (Scored 4).

Inoculation of tomato plants with *F. oxysporum* reduced AC but the effects were much greater when the ERM was disrupted before planting the tomato plants (Fig. 8.15). When the ERM was kept intact, the AC of tomato plants inoculated with the greatest concentration of *F. oxysporum* inoculum was not significantly different from the AC found in tomato plants not inoculated but the ERM had been disrupted, indicating that the ability of AMF to compete with *F. oxysporum* in colonizing the roots of the crop was enhanced

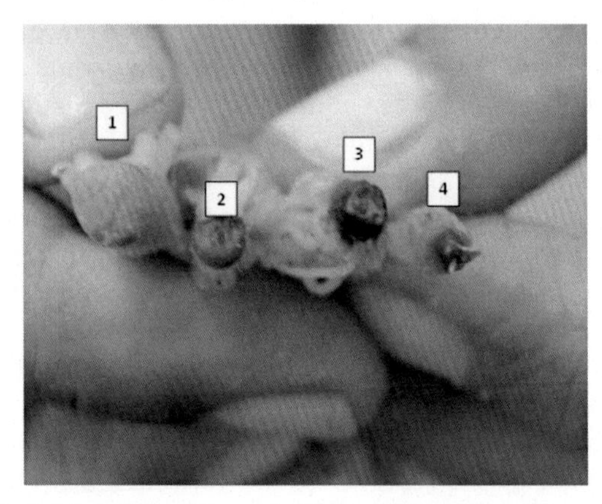


**PLATE 8.4** Grade for evaluation of disease incidence (DI) at the stem base of tomato.

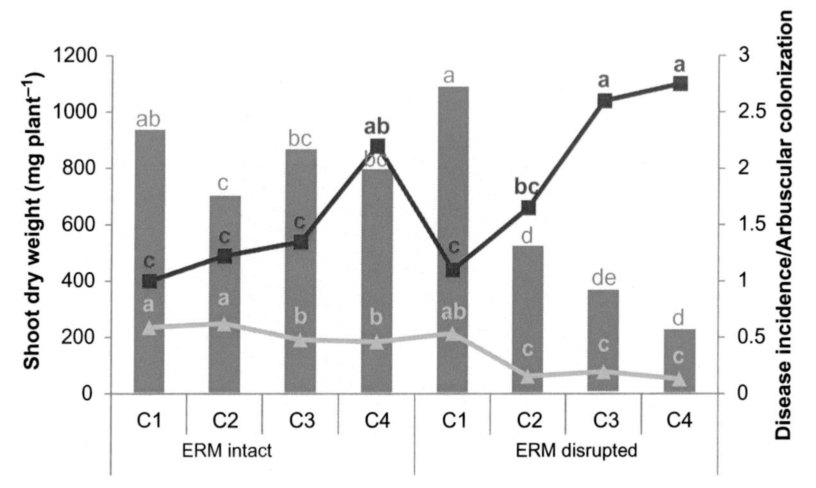


**FIGURE 8.15** Effect of the presence of an intact or disrupted extraradical mycelium (ERM) of arbuscular mycorrhizal fungi (AMF) at the time of tomato planting on the shoot dry weight (blue bars), disease incidence (DI) (red line and square markers) and arbuscular colonization (AC) (green line and triangular markers) 21 days after planting under inoculation of tomato plants with *Fusarium oxysporum* f.sp. *radicis-lycopersici* (C1, C2, C3, and C4 stands for inoculation with 0, $10^3$, $10^6$, and $10^9$ conidia plant$^{-1}$ at planting). ERM was previously developed in the soil by *Lolium rigidum* and indigenous AMF. For each parameter, the same letter indicates values are not significantly difference at $P = .05$.

by the presence of the intact ERM. The growth of tomato in the absence of *F. oxysporum* (C1 treatment) was not affected by the AMF inoculum source. However, when *F. oxysporum* was inoculated, the presence of intact ERM significantly increased the growth of tomato plants and reduced the DI (Fig. 8.15 and Plate 8.5). An intact ERM was critical for successful bioprotection of tomato against *F. oxysporum* and protection was achieved with the pathogenic fungus present in the roots of the crop at the time of planting.

A commercial operation producing field tomato reported a 6 ha visible occurrence of *F. oxysporum* in the Évora region of Portugal (Fig. 8.16). This resulted in a yield loss of about 20 tonnes ha$^{-1}$. Crop rotation was considered to be an unacceptable option for remediation, because it would require three consecutive years of tomato production to offset the costs of the specialized underground irrigation equipment. Consequently a field investigation was established to test the strategy considered in the pot experiment. In the field, that strategy was made practical by growing barley (*Hordeum vulgare* L.) as a mycotrophic cover crop over winter and keeping intact the ERM of the AM developed in association with its roots when the tomato crop was planted. The production system used by the company, which included deep tillage to loosen the soil compacted during harvest, was also modified. The normal production system used subsoiling, as the primary tillage, followed by disk harrowing and the formation of planting beds. All these operations

**PLATE 8.5** Tomato plants, 14 days old, inoculated with a suspension of $10^{-9}$ conidia of *Fusarium oxysporum* f.sp. *radicis-lycopersici* (inoculum level C4) at planting. Und and Dist are Undisturbed and Disturbed soil treatments described in Fig. 8.15. For Dist the soil was disturbed after the growth of *Lolium rigidum* G. (ERM disrupted) and in Und the soil remained undisturbed at the planting of tomato (ERM intact).

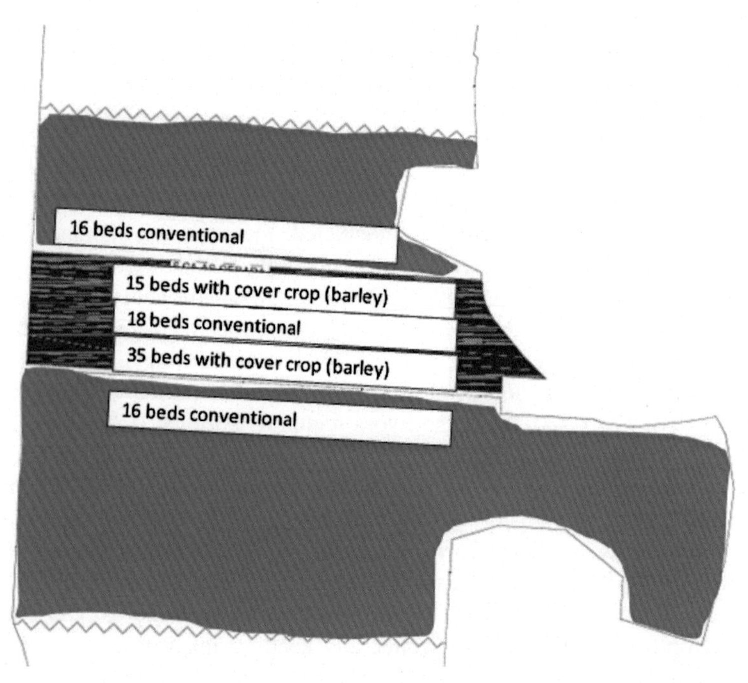


**FIGURE 8.16** Diagram of the field experiment with tomato. The red shading indicates the area (6 ha) where significant yield loss (20 $tonne^{-1}$) occurred in the 2013 season due to infection by *Fusarium oxysporum*. The green shading indicates the areas where no visible symptoms of the disease were identified and yield of the crop was considered normal for the region (100 tonnes $ha^{-1}$).

were carried out in spring, immediately before planting tomato. Such intensive tillage was incompatible with keeping intact the ERM developed in association with the barley. Therefore all essential tillage was performed as soon as possible after the harvest of the tomato crop. Barley was sown early in the winter after preparation of the beds and killed by systemic herbicide in the spring. Rebuilding the planting beds was achieved using a disk harrow operating to a depth of 10 cm just before the planting of tomato (Plate 8.6).

**PLATE 8.6** Field experiment with tomato. (A) Winter seeding of the barley cover crop after land preparation; (B) Barley cover crop in February; (C) Barley killed by systemic herbicide; (D) Traditional land preparation in spring before the planting of tomato.

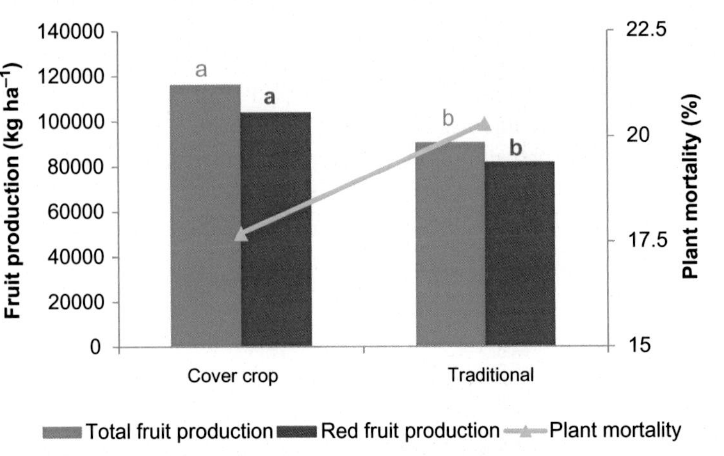


**FIGURE 8.17** Benefits for total and red fruit tomato production and reduced plant mortality from the switch to a winter cover crop (barley) with reduced tillage in spring, from the traditional practice. Results from a field experiment in the presence of *Fusarium oxysporum*. The cover crop treatment with minimum soil disturbance before the planting of tomato aimed to keep intact the ERM developed in association with barley roots; Traditional production system employed no cover crop but deep, and intensive tillage, including secondary disk harrowing in spring before planting the tomato. For each parameter the letter indicates values significantly difference at $P = .05$.

The benefit from the modified production system in maintaining an intact ERM developed between barley and indigenous mycorrhiza was significant (Fig. 8.17), even though all beds with the cover crop were located in the area where *F. oxysporum* had been most effective in reducing crop productivity in the previous year (Fig. 8.16). The yield obtained in the cropping system with the barley cover crop was considered exceptional for the region, even for fields where tomato had not been grown for several years. In both the laboratory experiment and the field investigation, Developer plants belonged to the Poaceae but for wider adoption it would be important to evaluate possible functional consequences of using ERM Developer plants from other botanical groups.

### 8.1.3 Discussion of the Results from the Case Studies

To understand the functional ecology of the AM symbiosis (see Sections 4.1 and 4.4) and how we can improve the effectiveness of indigenous AMF are crucial aspects to improve the relevance of mycorrhiza within the cropping systems (Miller et al., 1995). In all the results described in this chapter and a number of those discussed in others, indigenous mycorrhiza enhanced plant growth when an ERM, previously established in association with a Developer species, was kept intact at the planting of the test crop (see Chapter 6). In the presence of both abiotic (toxic levels of Mn) and biotic (*F. oxysporum*) stresses capable of impairing the growth of wheat, subterranean clover, and tomato, the improvement in crop performance was considerable but many benefits were lost if the inoculum available for colonization of a susceptible crop did not contain intact ERM. Therefore, the need for a well-established colonization by AMF before a plant encounters the stress, which has been considered a major limitation to the intentional use of AMF for bioprotection within agricultural systems (Sikora et al., 2008), can be overcome by promoting the presence of an intact ERM at the planting of a susceptible crop. Naturally when the benefit from AMF is the survival of the susceptible plants, the uptake of several nutrients, such as N, P, and K are also greatly enhanced. In establishing an intact ERM the botanical group of the Developer plant appears to determine the biological diversity of the AMF associated with the roots of the test crop. In the case of the field treatment of the biotic stressor, *F. oxysporum*, it was also important that the benefits achieved from the biocontrol were not prevented by a negative response of the mycorrhiza to the liberal application of nutrients. However, the negative effect of fertilizer application – within the range that plants respond to an application, especially the provision of P – on the colonization rate of host plants that is frequently referred to, may be more related to the method of assessment of AMF colonization than evidence of a negative effect on the ability of AMF to develop an intraradical mycelium and associated arbuscules. It may mainly be due to the effect of nutrient availability on the differential growth of the roots and fungus. The result obtained indicated that the strategy developed was applicable to a highly productive agricultural system, where no yield reduction, due to a limitation of the fertilizer application, would be acceptable to the grower.

Given all the evidence for functional diversity between different AMF species, and even between isolates of the same species (see Chapters 4 and 7), in relation to the benefits that these organisms can confer within an agricultural ecosystem, it seems likely that intentional management of the AMF diversity present in the roots of susceptible crops can be affected through the choice of Developer crop. This possibility might pave the way for increased advantages to be had from the use of AMF within agricultural systems. This seems to be the case with the significantly greater benefits for protection against Mn toxicity from ERM associated with an AMF consortium linked with *O. compressus* compared with one associated with *L. rigidum*, at least for wheat and subterranean clover, respectively (Brito et al., 2014; Alho et al., 2015; Brigido et al., 2017).

For nutrient acquisition by plants, the benefits from an intact ERM at the planting were also evident, even when large amounts of phosphorus were applied to the soil. The benefit of the presence of an intact ERM on P absorption is now well established in the literature, even when the earlier colonization of the crop is not enhanced (Miller et al., 1995). In addition to

P acquisition there was greater uptake of other important nutrients, including N, K, Ca, and S, even when application of P was compatible with large maize yields. These results were not confined to conditions where bioprotection was conferred by AMF against abiotic or biotic stresses able to kill the crop. The general strategy has been found to be compatible with high input cropping systems and therefore potentially a valuable contributor to improved fertilizer use efficiency under the concept of sustainable intensification. It therefore is also relevant to regions of the world where environmental impacts due to excessive use of fertilizers are of major concern.

*Production practices used within a cropping system must also need consider the role of AMF in sustainable intensification of agricultural systems.*

## 8.2 OPPORTUNITIES TO DEVELOP ERM FROM INDIGENOUS AMF WITHIN THE CROPPING SYSTEM

The key aspect of the general strategy we are advocating is the establishment of an ERM that acts as the dominant form of AMF propagule that can rapidly colonize the roots of plants as they emerge (Table 8.2). There are several opportunities within cropping systems to develop ERM from indigenous AMF in the soil. But a common element needs to be the adoption of appropriate tillage techniques to maintain the integrity of ERM, preparatory to planting the host crop that would benefit from mycorrhiza formation (Table 8.2).

The choice of crops within a rotation is an obvious aspect, considering that for the strategy the ERM needs to survive cold soil temperatures during the winter, typical of a temperate climate (McGonigle and Miller, 1999) or hot dry soils in summer, under Mediterranean environments (Jasper et al., 1989b; Brito et al., 2011). Cover crops provide another entry to enhance the role of AMF, particularly after nonmycotrophic elements in the crop rotation, such as oilseed rape (*Brassica napus* L.) or long bare fallows and hot temperatures (Kabir and Koide 2002; Lehman et al., 2012). However, the use of cover crops might be constrained under some cropping systems or environmental conditions. For instance under humid temperate conditions, such as in northern and central Europe, the gap between two consecutive winter crops may be too short to accommodate a cover crop. After oilseed rape or other nonmycotrophic crops, a possible solution could be to grow a mycotrophic cover crop over the fall and winter and sow a spring cereal. In the case of semiarid climates, there is no time to seed a cover crop between the beginnings of the rainy season and the sowing of the crop (Ehlers and Goss, 2016). If all the elements of the crop rotation are mycotrophic, considering that the ERM can survive to hot and dry periods that would probably be

**TABLE 8.2** Summary of the proposed strategy; mechanisms and benefits for the constructive management of arbuscular mycorrhizal fungi (AMF) within agricultural systems

| Strategy | | Mechanisms Exploited | Benefits |
|---|---|---|---|
| Cropping system supportive of AMF diversity and abundance | *Crop rotation*<br>*Cover crops*<br>*Rational use of input* | Ecological niches<br>Preferential associations between AMF and host plants<br>Reducing the levels of nutrients and pesticides in the soil | AMF population diverse and abundant |
| Compatible tillage system | *No-till*<br>*Strip tillage*<br>*Ridge-till*<br>*Reduced shallow cultivation* | Maintenance of ERM integrity | Earlier AM colonization to capitalize on the functional diversity of AMF associated with the Developer |
| User selected ERM Developer plant | *Crops in rotation;*<br>*Cover crops;*<br>*Weeds* | Preferential associations between AMF and host plant species | Managing AMF biological diversity |
| Crop to be protected | *Planting into **minimally** disturbed soil* | AMF colonization of crop plants initiated from an intact ERM<br>Functional diversity of AMF assemblage<br>Positive and negative feedback mechanisms | Enhancement of protection against biotic and abiotic stresses<br>Nutrient acquisition<br>Contribution of AMF to sustainable intensification |

enough to enhance the benefits of AMF. Alternatively mycotrophic weeds that germinate after the first rains could help to restore the ERM in the soil and increase the role of indigenous AMF in enhancing nutrient uptake and crop growth (Brito et al., 2013b).

In the driest part of sub-Saharan Africa and south Asia, transplanting techniques instead of direct seeding the crops have been studied as an opportunity to improve yields and ensure more uniform crops establishment with great success (Tenkouano et al., 1997; Murungu, et al., 2006; Adesina et al., 2014). The seeds are germinated in an irrigated nursery, which requires less water, before the beginning of the rainy season and transplanted to the field at the appropriate time. These are the regions of the world where the development of sustainable intensification of agriculture is more challenging, owing to economic and environmental constraints. The access of farmers to inputs, such as fertilizers, is very limited and abiotic stresses, such as water deficit, salinity or metalloid ions, and poor fertility are major constraints to soil productivity. Under these circumstances the results presented from the work developed by Brito et al. (2014) and Alho et al. (2015) could be a major contribution to improve the role of indigenous AMF in the bioprotection of the crops, the enhancement of nutrient uptake and improvement in yields. To put the proposed strategy into practice would only require small adjustments of the transplanting approach, which has already been tested and applied in those regions. Seedlings would have to be produced in individual containers (like bags), using soil from the fields where the crop is going to be planted. A mycotrophic Developer tolerant of the prevailing biotic or abiotic stressors would have to be sown first and grown for sufficient time for the mycorrhiza developed in association with its root system to form an ERM. The Developer would then need to be controlled by mechanically severing the shoot from the root system and without disrupting the soil before sowing the crop in the bag. The crop could then be planted after the beginning of the rainy season, already colonized by the AMF population originally associated with the Developer. Any tillage required in the field, for instance for weed control, would have to be performed before transplanting so that the ERM would remain intact. The crop plant would be well colonized by the time it came into contact with stressors and hence benefits conferred by mycorrhiza formation would be enhanced. The right choice of the Developer plant could improve the chances of getting functional advantages, as the AMF diversity associated with the roots of the Developer will be mainly the same as that colonizing the crop roots.

There remains the need to plant the saleable crop without disrupting the ERM established in relation to the Developer. Clearly this requires land preparation with minimum soil disturbance. Wherever and whenever possible a no-till operation would best meet this requirement. However, there are several circumstances where no-till is not compatible with the chosen cropping system. The most relevant examples would be the impossibility of using herbicides, including systems of organic agriculture or systems in regions with little or no access because of cost. Another circumstance would be associated with cropping systems, where currently soil compaction during harvest is impossible to avoid and deep tillage is required periodically to restore soil structure. The final group of difficult conditions pertains to rotations developed for regions of the world where cool soil temperatures in spring result in slow crop growth under no-till. Ridge or strip tillage have proved to be satisfactory alternatives as far as crop growth is concerned (Al-Kaisi et al., 2015) and the impact on AMF seems compatible with the proposed strategy. For example, McGonigle and Miller (1993) found that the earlier colonization of maize was greater under no-till and ridge-till than that under conventional tillage. For those situations where periodical remedial tillage is necessary to solve problems of soil compaction, the solution would be to anticipate the tillage operation and have time to sow a cover crop, thereby restoring an active ERM in the soil and use no-till or reduced tillage systems associated with herbicide applications to control the cover crop and weeds. In regions of the world where the farmers have no access to herbicides, tillage is normally performed by hand or animal traction and, therefore, the depth of tillage is normally superficial, which should be compatible with the proposed strategy. More difficult to address might be the case of organic farming. Research would be needed to develop tillage system that, as well as being effective for weed control, would also maintain the integrity of the ERM. Superficial soil tillage, timely and repeated in time might be compatible with the both objectives: weed control and the improvement of indigenous AMF performance within biological agricultural systems. The use of vigorous cover crops such as rye that are killed just before planting the crop has proved to be successful in controlling weeds preplanting, under no-till organic production (Shirtliffe and Johnson, 2012).

### 8.2.1 Criteria to Select Developer Plants

Prerequisite to manage AMF diversity and fully exploit the benefits of the mycorrhiza symbiosis is the maintenance of a diverse and abundant AMF community in the soil by adopting appropriate production techniques for all crops (see Chapter 2). However, present knowledge and resources are far away from allowing the determination of functional consequences associated with different AMF assemblages. The development of new molecular tools allows better understanding of AMF biodiversity but innovative approaches will be necessary to help the development of cropping systems targeted to the enhancement of functional benefits from diverse indigenous AMF populations (see Chapters 4 and 7).

To fully capitalize on the benefits from indigenous AMF will be decisive for managing the AMF diversity present in the root of the crops. This is because for each of the benefits from AM, functional diversity exists between different AM fungal species and even between isolates of the same species. There is plenty of information in the literature that report preferential associations between AMF and plant species. Therefore the AMF diversity present in the roots of Developer will depend on the plant functional group, which can be found at different levels of botanical affinity. But in a succession of two plants from different functional groups, if the colonization of the second plant is preferentially initiated by an intact ERM, the diversity of AMF present in its roots will have been influenced by the first plant in the sequence (Brígido et al., 2017) with functional consequences for protection of the second plant. Even when bioprotection is not the major role of AMF, there is evidence of functional consequences of managing the AMF diversity within a succession of different plant species. In an experiment where the AMF population was influenced by two consecutive generations of two different host plants, *Plantago lanceolata* and *Panicum sphaerocarpon*, Bever (2002) showed evidences of negative feedback mechanisms because the AMF that accumulate in the soil after the growth of *Panicum* were better at promoting the growth of *Plantago* than were the AMF accumulated by the *Plantago* itself.

When abiotic stresses, such as excessive levels of salt or metalloid ions, water stress or very limited soil contents of P and other less mobile nutrients, cause major limitations to crop performance, it is reasonable to assume that indigenous AMF will be adapted to the conditions. Appropriate functional diversity within the population is also likely to be found. The problem will be to select the Developers that will harbor the AMF that are more efficient in protecting the economically important plants. A sensible approach would be to choose the Developer plant from the naturally occurring vegetation in the area. Preselection criteria could be to test the tolerance of those plants with and without the presence of the indigenous AMF, i.e., with and without soil sterilization. From the two ERM Developer plant species used in the work on protection from Mn toxicity, *O. compressus* provided significantly greater bioprotection than *L. rigidum*. In a subsequent study, which required sterilization of the soil that contained excessive levels of Mn, *O. compressus* was not able to grow, whereas *L. rigidum* still grew vigorously. The observation was consistent with *O. compressus* being more dependent on the symbiosis for its protection against Mn toxicity. Therefore, it seems reasonable to suggest that selecting Developer plants that are only tolerant to the stress in the presence of indigenous AMF would increase the chances of identifying ones that are more efficient in protecting other plants against the stressor. Another indication of the greater degree of dependency on AM by *O. compressus* is the size of its root system. After 7 weeks growth the weight of the root system of *L. rigidum* was almost the double that of *O. compressus* (Brito et al., 2014). This morphological criterion might be also important when the role of the developer is to improve the uptake of nutrients, because plants that do not normally develop dense or hairy root systems are more dependent on the symbiosis for the absorption of less mobile nutrients (Baylis, 1972; Schweiger et al., 1995). Considering that the benefits of AM depend on both partners of the symbiosis (Hart et al., 2013), selecting the best Developer will have to consider the stresses to be overcome and the crop to be protected, bearing in mind the existence of negative feedback mechanisms (Bever, 2002).

For protection against biotic stresses, particularly soilborne diseases, the selection of the best Developers might be more arduous since the plant must not host the pathogen. Therefore, the degree of the dependency of the Developer plant on its association with AMF for bioprotection cannot be used as preselection criteria. A broad screening process may be the only way of selecting Developers.

## 8.3 SOME FINAL COMMENTS

An important part of the strategy outlined in this chapter is the establishment of crop sequences, either within a rotation or through the use of cover crops. Within cropping systems there is an urgent need for research to clarify the mechanisms underpinning the positive interactions between different crops. In reviewing the literature, Dias et al. (2015) stated that they "could not find a single (paper) containing science-based criteria for crop rotations." Interaction between different elements of the crop rotation and soil microbes are certainly involved and feedback mechanisms with AMF, positive or negative, must be investigated.

In considering the challenge of how to feed the growing world population in the near future but reducing the environmental impacts of agriculture, particularly soil degradation, water conservation, energy efficiency, and loss of biodiversity we developed the framework (Table 8.2) that resulted in our proposed strategy. The possibility we now have to manage indigenous AMF provides a rationale for increasing the role of mycorrhiza in the development of a sustainable intensification of agricultural systems in different parts of the world. As we have indicated the components of a possible system for managing AMF are slowly coming together, which could make its implementation in commercial agriculture an imminent possibility. Clearly research requirements are numerous but modern genomics technology is bringing closer the possibilities for identifying compatible assemblages of AMF and helper bacteria. The establishment of available molecular markers in host plants, which would allow confirmation of the presence of an effective mycorrhiza, still has to take place. The importance of host preference rather than specificity in AMF permits the use of common weed species and different elements of crop rotation to act as ERM Developer plants but potentially hinders the selection of effective AMF assemblages. But whether members of the Fabaceae are more effective developers than other families still has to be determined. Issues over land preparation including tillage, weed control, nutrient management, and crop selection still need fine tuning if we are to optimize conditions for rapid and early colonization of target hosts with effective AMF. Greater emphasis on strategic weed, disease, and pest control and judicious use of mineral fertilizers is likely.

# References

Abbaspour, H., Saeidi-Sar, S., Afshari, H., Abdel-Wahhab, M.A., 2012. Tolerance of mycorrhiza infected Pistachio (*Pistacia vera* L.) seedling to drought stress under glasshouse conditions. J. Plant Physiol. 169, 704–709.

Abbott, L.K., Robson, A.D., 1981. Infectivity and effectiveness of vesicular arbuscular mycorrhizal fungi: effect of inoculum type. Aust. J. Agric. Res. 32, 631–639.

Abbott, L.K., Robson, A.D., 1982. Infectivity of vesicular arbuscular mycorrhizal fungi in agricultural soils. Aust. J. Agric. Res. 33, 1049–1059.

Abbott, L.K., Robson, A.D., 1985a. Formation of external hyphae in soil by four species of vesicular-arbuscular mycorrhizal fungi. New Phytol. 99, 245–255.

Abbott, L.K., Robson, A.D., 1985b. The effect of soil pH on the formation of VA mycorrhizas by two species of Glomus. Aust. J. Soil Res. 23, 253–261.

Abbott, L.K., Robson, A.D., 1991. Field management of VA mycorrhizal fungi. In: Keister, D.L., Cregan, P.B. (Eds.), The Rhizosphere and Plant Growth. Kluwer Academic Publishers, Dordrecht, Netherlands, pp. 355–362.

Abbott, L.K., Robson, A.D., Gazey, C., 1992. Selection of inoculant vesicular-arbuscular mycorrhizal fungi. In: Norris, J.R., Read, D.J., Varma, A.K. (Eds.), Methods in Microbiology, vol. 24. Academic Press, London, pp. 1–21.

Abbott, L.K., Robson, A.D., Scheltema, M.A., 1995. Managing soils to enhance mycorrhizal benefits in Mediterranean agriculture. Crit. Rev. Biotechnol. 15, 213–228.

Abdalla, M.E., Abdel-Fattah, G.M., 2000. Influence of the endomycorrhizal fungus *Glomus mosseae* on the development of peanut pod rot disease in Egypt. Mycorrhiza 10, 29–35.

Abd-Allaa, M.H., Omara, S.A., Karanxha, S., 2000. The impact of pesticides on arbuscular mycorrhizal and nitrogen-fixing symbioses in legumes. Appl. Soil Ecol. 14, 191–200.

Achatz, M., Rillig, M.C., 2014. Arbuscular mycorrhizal fungal hyphae enhance the transport of the allelochemical juglone in the field. Soil Biol. Biochem. 78, 76–82.

Addy, H.D., Boswell, E.P., Koide, R.T., 1998. Low temperature acclimation and freezing resistance of extraradical VA mycorrhizal hyphae. Mycol. Res. 102, 582–586.

Addy, H.D., Miller, M.H., Peterson, R.L., 1997. Infectivity of the propagules associated with extraradical mycelia of two AM fungi following winter freezing. New Phytol. 135, 745–753.

Adesina, J.M., Afolabi, L.A., Aderibigbe, A.T.B., Sanni, K.O., 2014. Effect of direct seeding and transplanting on the performance of maize (*Zea mays* L.) in South-Western Nigeria. Researcher 6, 16–20.

Adholeya, A., Tiwari, P., Singh, R., 2005. Large-scale inoculum production of arbuscular mycorrhizal fungi on root organs and inoculation strategies. In: Declerck, S., Strullu, D.-G., Fortin, A. (Eds.), In Vitro Culture of Mycorrhizas. Springer-Verlag, Berlin and Heidelberg, pp. 315–338. Soil Biology, vol. 4, Part 2.

Adolfsson, L., Solymosi, K., Andersson, M.X., Keresztes, Á., Uddling, J., Schoefs, B., et al., 2015. Mycorrhiza symbiosis increases the surface for sunlight capture in *Medicago truncatula* for better photosynthetic production. PLoS ONE 10 (1), e0115314. Available from: http://dx.doi.org/10.1371/journal.pone.0115314.

Ahmed, F.R.S., Killham, K., Alexander, I., 2006. Influences of arbuscular mycorrhizal fungus *Glomus mosseae* on growth and nutrition of lentil irrigated with arsenic contaminated water. Plant Soil 283, 33–41.

Akhtar, M.S., Siddiqui, Z.A., 2008. *Glomus intraradices*, *Pseudomonas alcaligenes*, and *Bacillus pumilus*: effective agents for the control of root-rot disease complex of chickpea (*Cicer arietinum* L.). J. Gen. Plant Pathol. 74, 53–60.

Akiyama, K., Matsuzaki, K.-I., Hayashi, H., 2005. Plant sesquiterpenes induce hyphal branching in arbuscular mycorrhizal fungi. Nature 435, 824–827.

Albertsen, A., Ravnskov, S., Green, H., Jensen, D.F., Larsen, J., 2006. Interactions between mycelium of the mycorrhizal fungus *Glomus intraradices* and other soil microorganisms as affected by organic matter. Soil Biol. Biochem. 38, 1008–1014.

Alexandratos, N., Bruinsma, J., 2012. World agriculture towards 2030/2050. The 2012 Revision. ESA Working Paper No. 12-03. FAO, Food and Agriculture Organization of the United Nations, Rome. Available from: <http://www.fao.org/docrep/016/ap106e/ap106e.pdf> (accessed 08.12.15).

Al-Garni, S.M.S., 2006. Increasing NaCI-salt tolerance of a halophytic plant *Phragmites australis* by mycorrhizal symbiosis. Am.-Eurasian J. Agric. Environ. Sci. 1, 119–126.

Alguacil, M.D., Lozano, Z., Campoy, M.J., Roldan, A., 2010. Phosphorus fertilisation management modifies the biodiversity of AM fungi in a tropical savanna forage system. Soil Biol. Biochem. 42, 1114–1122.

Alguacil, M.M., Lumini, E., Roldan, A., Salinas-Garcia, J.R., Bonfante, P., Bianciotto, V., 2008. The impact of tillage practices on arbuscular mycorrhizal fungal diversity in subtropical crops. Ecol. Appl. 18, 527–536.

Alguacil, M.M., Torrecillas, E., García-Orenes, F., Roldán, A., 2014. Changes in the composition and diversity of AMF communities mediated by management practices in a Mediterranean soil are related with increases in soil biological activity. Soil Biol. Biochem. 76, 34–44.

Alho, L., Carvalho, M., Brito, I., Goss, M.J., 2015. The effect of arbuscular mycorrhiza fungal propagules on the growth of subterranean clover (*Trifolium subterraneum* L.) under Mn toxicity in ex situ experiments. Soil Use Manage. 31, 337–344.

Al-Kaisi, M.M., Archontoulis, S.V., Kwaw-Mensah, D., Miguez, F., 2015. Tillage and crop rotation effects on corn agronomic response and economic return at seven Iowa locations. Agron. J. 107, 1411–1424.

Alkan, N., Gadkar, V., Yarden, O., Kapulnik, Y., 2006. Analysis of quantitative interactions between two species of arbuscular mycorrhizal fungi, *Glomus mosseae* and G. intraradices, by real-time PCR. Appl. Environ. Microbiol. 72, 4192–4199. Available from: http://dx.doi.org/10.1128/AEM.02889-05.

Al-Karaki, G.N., Clark, R.B., 1998. Growth mineral acquisition and water use by mycorrhizal wheat grown under water stress. J. Plant Nutr. 21, 263–276.

Allen, M.F., 2007. Mycorrhizal fungi: highways for water and nutrients in arid soils. Vadose Zone J. 6, 291–297.

Alonso, L.M., Kleiner, D., Ortega, E., 2008. Spores of the mycorrhizal fungus *Glomus mosseae* host yeasts that solubilize phosphate and accumulate polyphosphates. Mycorrhiza 18, 197–204.

Ames, R.N., Reid, C.P.P., Porter, L.K., Cambardella, C., 1983. Hyphal uptake and transport of nitrogen from 2 n-15-labeled sources by *Glomus mosseae*, a vesicular arbuscular mycorrhizal fungus. New Phytol. 95, 381–396.

Amijee, F., Tinker, P.B., Stribley, D.P., 1989. The development of endomycorrhizal root systems. New Phytol. 111, 435–446.

An, Z.Q., Hendrix, J.W., Hershman, D.E., Ferriss, R.S., Henson, G.T., 1993. The influence of crop rotation and soil fumigation on a mycorrhizal fungal community associated with soybean. Mycorrhiza 3, 171–182.

Anasiru, R.H., Rayes, M.L., Setiawan, B., Soemarno, 2013. Economic valuation of soil erosion on cultivated drylands in Langge sub watershed, Gorontalo, Indonesia. J. Nat. Sci. Res. 3, 40–48, www.iiste.org ISSN 2224–3186.

Andrade, G., Mihara, K.L., Linderman, R.G., Bethlenfalvay, G.J., 1997. Bacteria from rhizosphere and hyphosphere soils of different arbuscular-mycorrhizal fungi. Plant Soil 192, 71–79.

Andrade, G., Mihara, K.L., Linderman, R.G., Bethlenfalvay, G.J., 1998. Soil aggregation status and rhizobacteria in the mycorrhizosphere. Plant Soil 202, 89–96.

Angelard, C., Colard, A., Niculita-Hirzel, H., Croll, D., Sanders, I.R., 2010. Segregation in a mycorrhizal fungus alters rice growth and symbiosis-specific gene transcription. Curr. Biol. 20, 1216–1221.

Antoninka, A., Reich, P.B., Johnson, N.C., 2011. Seven years of carbon dioxide enrichment, nitrogen fertilization and plant diversity influence arbuscular mycorrhizal fungi in a grassland ecosystem. New Phytol. 192, 200–214.

Antoninka, A., Wolf, J.E., Bowker, M., Classen, A.T., Johnson, N.C., 2009. Linking above and belowground responses to global change at community and ecosystem scales. Global Change Biol. 15, 914–929.

Antunes, P.M., de Varennes, A., Rajcan, I., Goss, M.J., 2006a. Accumulation of specific flavonoids in soybean (*Glycine max* (L.) Merr.) as a function of the early tripartite symbiosis with arbuscular mycorrhizal fungi and *Bradyrhizobium japonicum* (Kirchner) Jordan. Soil Biol. Biochem. 38, 1234–1242.

Antunes, P.M., de Varennes, A., Zhang, T., Goss, M.J., 2006b. The tripartite symbiosis formed by indigenous arbuscular mycorrhizal fungi, *Bradyrhizobium japonicum* and soya bean under field conditions. J. Agron. Crop Sci. 192, 373–378.

Antunes, P.M., Schneider, K., Hillis, D., Klironomos, J.N., 2007. Can the arbuscular mycorrhizal fungus *Glomus intraradices* actively mobilize P from rock phosphates? Pedobiologia 51, 281–286.

Antunes, P.M., Koch, A.M., Morton, J.B., Rillig, M.C., Klironomos, J.N., 2011. Evidence for functional divergence in arbuscular mycorrhizal fungi from contrasting climatic origins. New Phytol. 189, 507–514.

Antunes, P.M., Lehmann, A., Hart, M.M., Baumecker, M., Rillig, M.C., 2012. Long-term effects of soil nutrient deficiency on arbuscular mycorrhizal communities. Funct. Ecol. 26, 532–540.

Arihara, J., Karasawa, T., 2000. Effect of previous crops on arbuscular mycorrhizal formation and growth of succeeding maize. Soil Sci. Plant Nutr. 46, 43–51.

Artursson, V., Finlay, R.D., Jansson, J.K., 2006. Interactions between arbuscular mycorrhizal fungi and bacteria and their potential for stimulating plant growth. Environ. Microbiol. 8, 1–10.

Aryal, U.K., Shah, S.K., Xu, H.L., Fujita, M., 2007. Growth, nodulation and mycorrhizal colonization in bean plants improved by rhizobial inoculation with organic and chemical fertilization. J. Sustainable Agric. 29, 71–83.

Asghari, H.R., Cavagnaro, T.R., 2011. Arbuscular mycorrhizas enhance plant interception of leached nutrients. Funct. Plant Biol. 38, 219–226.

Atkinson, D., 2004. Low input agriculture: the role of arbuscular mycorrhizal fungi. In: Baar, J., Josten, E. (Eds.), Role of Mycorrhiza in Sustainable Land Management. Programme and abstracts of the COST-Meeting: Role of Mycorrhiza in Sustainable Land Management. Applied Plant Research, WUR, Netherlands.

Atkinson, D., 2009. Soil microbial resources and agricultural policies. In: Azcón-Aguilar, C., Barea, J.M., Gianinazzi, S., Gianinazzi-Pearson, V. (Eds.), Mycorrhizas Functional Processes and Ecological Impact. Springer-Verlag, Berlín, Heidelberg, pp. 33–45.

Audet, P., Charest, C., 2006. Effects of AM colonization on "wild tobacco" plants grown in zinc contaminated soil. Mycorrhiza 16, 277–283.

Augé, R.M., 2001. Water relations, drought and vesicular-arbuscular mycorrhizal symbiosis. Mycorrhiza 11, 3–42.

Augé, R.M., 2004. Arbuscular mycorrhizae and soil/plant water relations. Can. J. Soil Sci. 84, 373–381.

Augé, R.M., Moore, J.L., Cho, K., Stutz, J.C., Sylvia, D.M., Al-Agely, A., et al., 2003. Relating dehydration resistance of mycorrhizal *Phaseolus vulgaris* to soil and root colonization by hyphae. J. Plant Physiol. 160, 1147–1156.

Augé, R.M., Stodola, A.J.W., Tims, J.E., Saxton, A.M., 2001. Moisture retention properties of a mycorrhizal soil. Plant Soil 230, 87–97.

Augé, R.M., Toler, H.D., Saxton, A.M., 2015. Arbuscular mycorrhizal symbiosis alters stomatal conductance of host plants more under drought than under amply watered conditions: a meta-analysis. Mycorrhiza 25, 13–24.

Avio, L., Castaldini, M., Fabiani, A., Bedini, S., Sbrana, C., Turrini, A., et al., 2013. Impact of nitrogen fertilization and soil tillage on arbuscular mycorrhizal fungal communities in a Mediterranean agroecosystem. Soil Biol. Biochem. 67, 285–294.

Avio, L., Pellegrino, E., Bonari, E., Giovannetti, M., 2006. Functional diversity of arbuscular mycorrhizal fungal isolates in relation to extraradical mycelial networks. New Phytol. 172, 347–357.

Azcón, R., Barea, J.M., 2010. Mycorrhizosphere interactions for legume improvement. In: Khan, M.S., Zaidi, A., Musarrat, J. (Eds.), Microbes for Legume Improvement. Springer-Verlag, Heidelberg, pp. 237–271.

Azcón, R., Rubio, R., Barea, J.M., 1991. Selective interactions between different species of mycorrhizal fungi and *Rhizobium meliloti* strains, and their effects on growth, $N_2$-fixation ($^{15}N$) and nutrition of *Medicago sativa* L. New Phytol. 117, 399–404.

Azcón, R., Ruíz-Lozano, J.M., Rodríguez, R., 2001. Differential contribution of arbuscular mycorrhizal fungi to plant nitrate uptake ($N^{15}$) under increasing N supply to the soil. Can. J. Bot. 79, 1175–1180.

Azcón-Aguilar, C., Barea, J.M., 2015. Nutrient cycling in the mycorrhizosphere. J. Soil Sci. Plant Nutr. 25, 372–396.

Azcón-Aguilar, C., Barea, J.M., Olivares, J., 1980. Effects of Rhizobium polysaccharides on VA mycorrhiza formation. Second International Symposium on Microbial Ecology. University of Warwick, Coventry, Abstract No 187.

Babikova, Z., Gilbert, L., Bruce, T., Dewhirst, S.Y., Pickett, J.A., Johnson, D., 2013. Arbuscular mycorrhizal fungi and aphids interact by changing host plant quality and volatile emission. Ecol. Lett. 16, 835–843.

Bago, B., Azcon–Aguilar, C., Piché, Y., 1998. Architecture and developmental dynamics of the external mycelium of the arbuscular mycorrhizal fungus *Glomus intraradices* grown under monoxenic conditions. Mycologia 90, 52–62.

Bakonyi, G., Posta, K., Kiss, I., Fábián, M., Nagy, P., Nosel, J.N., 2002. Density-dependent regulation of arbuscular mycorrhiza by collembola. Soil Biol. Biochem. 34, 661–664.

Baltruschat, H., Dehne, H.W., 1989. The occurrence of vesicular-arbuscular mycorrhiza in agroecosystems II. Influence of nitrogen fertilization and green manure in continuous monoculture and in crop rotation on the inoculum potential of winter barley. Plant Soil 113, 251–256.

Bamforth, S.S., 1971. The numbers and proportions of Testacea and ciliates in litters and soils. J. Protozool. 18, 24–28.

Barbieri, P.A., Rozas, H.R.S., Covacevich, F., Echeverría, H.E., 2014. Phosphorus placement effects on phosphorous recovery efficiency and grain yield of wheat under no-tillage in the humid pampas of Argentina. Int. J. Agron. Article ID 507105, 12 p. http://dx.doi.org/10.1155/2014/507105.

Bardgett, R.D., van der Putten, W.H., 2014. Belowground biodiversity and ecosystem functioning. Nature 515, 505–511.

Bardgett, R.D., Wardle, D.A., 2010. Aboveground-Belowground Linkages: Biotic Interactions, Ecosystem Processes, and Global Change. Oxford University, Oxford.

Barea, J.M., Palenzuela, J., Cornejo, P., Sanchez-Castro, I., Navarro-Fernandez, C., Lopez-Garcia, A., et al., 2011. Ecological and functional roles of mycorrhizas in semi-arid ecosystems of southeast Spain. J. Arid Environ. 75, 1292–1301.

Barros, J.F.C., Basch, G., Carvalho, M., 2008. Effect of reduced doses of a post-emergence graminicide to control *Avena sterilis* L. and *Lolium rigidum* G. in no-till wheat under Mediterranean environment. Crop Prot. 27, 1031–1037.

Barto, E.K., Weidenhamer, J.D., Cipollini, D., Rillig, M.C., 2012. Fungal superhighways: Do common mycorrhizal networks enhance below ground communication? Trends Plant Sci. 17, 633–637.

Bates, S., Berg-Lyons, D., Caporaso, J.G., Walters, W.A., Knight, R., Fierer, N., 2011. Examining the global distribution of dominant archaeal populations in soil. ISME J. 5, 908–917.

Baylis, G.T.S., 1972. Minimum levels of available phosphorus for non-mycorrhizal plants. Plant Soil 36, 233–234.

Beckie, H.J., 2006. Herbicide-resistant weeds: management tactics and practices. Weed Technol. 20, 793–814.

Becklin, K.M., Hertweck, K.L., Jumpponen, A., 2012. Host identity impacts rhizosphere fungal communities associated with three alpine plant species. Microb. Ecol. 63, 682–693.

Bedini, S., Avio, L., Sbrana, C., Turrini, A., Migliorini, P., Vazzana, C., et al., 2013. Mycorrhizal activity and diversity in a long-term organic Mediterranean agroecosystem. Biol. Fertil. Soils 49, 781–790.

Bedini, S., Pellegrino, E., Avio, L., Pellegrini, S., Bazzoffi, P., Argese, E., et al., 2009. Changes in soil aggregation and glomalin-related soil protein content as affected by the arbuscular mycorrhizal fungal species *Glomus mosseae* and *Glomus intraradices*. Soil Biol. Biochem. 41, 1491–1496.

Bender, S.F., Conen, F., Van der Heijden, M.G.A., 2015. Mycorrhizal effects on nutrient cycling, nutrient leaching and $N_2O$ production in experimental grassland. Soil Biol. Biochem. 80, 283–292.

Bender, S.F., van der Heijden, M.G.A., 2015. Soil biota enhance agricultural sustainability by improving crop yield, nutrient uptake and reducing nitrogen leaching losses. J. Appl. Ecol. 52, 228–239.

Berendsen, R.L., Pieterse, C.M.J., Bakker, P., 2012. The rhizosphere microbiome and plant health. Trends Plant Sci. 17, 478–486.

Besserer, A., Puech-Pagès, V., Kiefer, P., Gomez-Roldan, V., Jauneau, A., Roy, S., et al., 2006. Strigolactones stimulate arbuscular mycorrhizal fungi by activating mitochondria. PLoS Biol. 4, 1239–1247.

Bethlenfalvay, G.J., Brown, M.S., Stafford, A.E., 1985. The Glycine-Glomus-Rhizobium symbiosis. II. Antagonistic effects between mycorrhizal colonization and nodulation. Plant Physiol. 79, 1054–1058.

Bethlenfalvay, G.J., Pacovsky, R.S., Bayne, H.G., Stafford, A.E., 1982. Interactions between nitrogen fixation, mycorrhizal colonization and host-plant growth in the *Phaseolus-Rhizobium-Glomus* symbiosis. Plant Physiol. 70, 446–450.

Bethlenfalvay, G.J., Reyes–Solis, M.G., Camel, S.B., Ferrera–Cerrato, R., 1991. Nutrient transfer between the root zones of soybean and maize plants connected by a common mycorrhizal mycelium. Physiol. Plant 82, 423–432.

Bever, J., 2002. Negative feedback within a mutualism: host-specific growth of mycorrhizal fungi reduces plant benefit. Proc. R. Lond. B 269, 2595–2602.

Bever, J.D., 2003. Soil community feedback and the coexistence of competitors: conceptual frameworks and empirical tests. New Phytol. 157, 465–473.

Bever, J.D., Richardson, S.C., Lawrence, B.M., Holmes, J., Watson, M., 2009. Preferential allocation to beneficial symbiont with spatial structure maintains mycorrhizal mutualism. Ecol. Lett. 12, 13–21.

Bhadalung, N.N., Suwanarit, A., Dell, B., Nopamornbod, O., Thamchaipenet, A., Rungchuang, J., 2005. Effects of long-term NP-fertilization on abundance and diversity of arbuscular mycorrhizal fungi under a maize cropping system. Plant Soil 270, 371–382.

Bianciotto, V., Bandi, C., Minerdi, D., Sironi, M., Tichy, H.V., Bonfante, P., 1996. An obligately endosymbiotic fungus itself harbors obligately intracellular bacteria. Appl. Environ. Microbiol. 62, 3005–3010.

Bianciotto, V., Genre, A., Jargeat, P., Lumini, E., Bécard, G., Bonfante, P., 2004. Vertical transmission of endobacteria in the arbuscular mycorrhizal fungus *Gigaspora margarita* through generation of vegetative spores. Appl. Environ. Microbiol. 70, 3600–3608.

Biermann, B., Linderman, R.G., 1983. Use of vesicular arbuscular mycorrhizal roots, intraradical vesicles and extraradical vesicles as inoculum. New Phytol. 95, 97–105.

Bilalis, D.J., Karamanos, A.J., 2010. Organic maize growth and mycorrhizal root colonization response to tillage and organic fertilization. J. Sustainable Agric. 34, 836–849.

Binaj, A., Veizi, P., Beqiraj, E., Gjoka, F., Kasa, E., 2014. Economic losses from soil degradation in agricultural area in Albania. Agric. Econ. Czech 60, 287–293.

Black, R., Tinker, P.B., 1979. Development of endomycorrhizal root systems. 2. Effect of agronomic factors and soil-conditions on the development of vesicular-arbuscular mycorrhizal infection in barley and on the endophyte spore density. New Phytol. 83, 401–413.

Blanco-Canqui, H., Stone, L.R., Schlegal, A.J., Benjamin, J.G., Vigil, M.F., Stahlman, P.W., 2010. Continuous cropping systems reduce near-surface maximum compaction in no-till soils. Agron. J. 102, 1217–1225.

Blanke, V., Renker, C., Wagner, M., Füllner, K., Held, M., Kuhn, A.J., et al., 2005. Nitrogen supply affects arbuscular mycorrhizal colonization of *Artemisia vulgaris* in a phosphate polluted field site. New Phytol. 166, 981–992.

Blume, E., Bischoff, M., Reichert, J.M., Moorman, T., Konopka, A., Turco, R.F., 2002. Surface and subsurface microbial biomass, community structure and metabolic activity as a function of soil depth and season. Appl. Soil Ecol. 20, 171–181.

Boby, V.U., Balakrishna, A.N., Bagyaraj, D.J., 2008. Interaction between *Glomus mosseae* and soil yeasts on growth and nutrition of cowpea. Microbiol. Res. 163, 693–700.

Boddington, C.L., Bassett, E.E., Jakobsen, I., Dodd, J.C., 1999. Comparison of techniques for the extraction and quantification of extra–radical mycelium of arbuscular mycorrhizal fungi in soils. Soil Biol. Biochem. 31, 479–482.

Bodelier, P.L.E., 2011. Toward understanding, managing, and protecting microbial ecosystems. Front. Microbiol. 2, 80.

Bolan, N., 1991. A critical review on the role of mycorrhizal fungi in the uptake of phosphorus by plants. Plant Soil 134, 189–207.

Bompadre, M.J., Molina, M.C.R., Colombo, R.P., Bidondo, L.F., Silvani, V.A., Pardo, A.G., et al., 2013. Differential efficiency of two strains of the arbuscular mycorrhizal fungus *Rhizophagus irregularis* on olive (*Olea europaea*) plants under two water regimes. Symbiosis 61, 105–112.

Bonfante, P., Anca, I.-A., 2009. Plants, mycorrhizal fungi, and bacteria: a network of interactions. Annu. Rev. Microbiol. 63, 363–383.

Bonfante-Fasolo, P., Perotto, S., 1992. Plant and endomycorrhizal fungi: the cellular and molecular basis of their interaction. In: Verma, D.P.S. (Ed.), Molecular Signals in Plant–Microbe Communications. CRC Press, Boca Raton, FL, pp. 445–470.

Borowicz, V.A., 2001. Do arbuscular mycorrhizal fungi alter plant-pathogen relations? Ecology 82, 3057–3068.

Borriello, R., Lumini, E., Girlanda, M., Bonfante, P., Bianciotto, V., 2012. Effects of different management practices on arbuscular mycorrhizal fungal diversity in maize fields by a molecular approach. Biol. Fertil. Soils 48, 911–922.

Börstler, B., Raab, P.A., Thiéry, O., Morton, J.B., Redecker, D., 2008. Genetic diversity of the arbuscular mycorrhizal fungus Glomus intraradices as determined by mitochondrial large subunit rRNA gene sequences is considerably higher than previously expected. New Phytol. 180, 452–465. Available from: http://dx.doi.org/10.1111/j.1469-8137.2008.02574.x.

Börstler, B., Thiéry, O., Sykorová, Z., Berner, A., Redecker, D., 2010. Diversity of mitochondrial large subunit rDNA haplotypes of Glomus intraradices in two agricultural field experiments and two semi-natural grasslands. Mol. Ecol. 19, 1497–1511. Available from: http://dx.doi.org/10.1111/j.1365-294X.2010.04590.x.

Bothe, H., 2012. Arbuscular mycorrhiza and salt tolerance of plants. Symbiosis 58, 7–16.

Bouffaud, M.-L., Bernaud, E., Colombet, A., van Tuinen, D., Wipf, D., Redecker, D., 2015. Regional-scale analysis of arbuscular mycorrhizal fungi: the case of Burgundy vineyards. Journal International des Sciences de la Vigne et du Vin 49, 275–291, doi:10.1016/s0038-0717(97)00259-9.

Bowen, G.D., Rovira, A.D., 1999. The rhizosphere and its management to improve plant growth. Adv. Agron. 66, 1–102.

Boyetchko, S.M., Tewari, J.P., 1995. Susceptibility of barley cultivars to vesicular-arbuscular mycorrhizal fungi. Can. J. Plant Sci. 75, 269–275.

Bradley, K., Drijber, R.A., Knops, J., 2006. Increased N availability in grassland soils modifies their microbial communities and decreases the abundance of arbuscular mycorrhizal fungi. Soil Biol. Biochem. 38, 1583–1595.

Braunberger, P.G., Abbott, L.K., Robson, A.D., 1996. Infectivity of arbuscular mycorrhizal fungi after wetting and drying. New Phytol. 134, 673–684.

Braunberger, P.G., Miller, M.H., Peterson, R.L., 1991. Effect of phosphorus nutrition on morphological characteristics of vesicular-arbuscular colonization of maize. New Phytol. 119, 107–113.

Bressan, M., Trinsoutrot Gattin, I., Desaire, S., Castel, L., Gangneux, C., Laval, K., 2015. A rapid flow cytometry method to assess bacterial abundance in agricultural soil. Appl. Soil Ecol. 88, 60–68.

Brígido, C., van Tuinen, D., Brito, I., Alho, L., Goss, M.J., Carvalho, M., 2017. Management of the biological diversity of AM fungi by combination of host plant succession and integrity of extraradical mycelium. Soil Biol. Biochem. (In Press).

Brito, I., De Carvalho, M., Goss, M.J., 2009. Techniques for arbuscular mycorrhiza inoculum reduction. In: Varma, A., Kharkwal, A.C. (Eds.), Symbiotic Fungi, Soil Biology, vol. 18. Springer-Verlag, Berlin Heidelberg, pp. 307–319.

Brito, I., De Carvalho, M., Goss, M.J., 2011. Summer survival of arbuscular mycorrhiza extraradical mycelium and the potential for its management through tillage options in Mediterranean cropping systems. Soil Use Manage. 27, 350–356.

Brito, I., Goss, M.J., Carvalho, M., 2012a. Effect of tillage and crop on arbuscular mycorrhiza colonisation of winter wheat and triticale under Mediterranean conditions. Soil Use Manage. 28, 202–208.

Brito, I., Goss, M.J., Carvalho, M., Chatagnier, O., Van Tuinen, D., 2012b. Impact of tillage system on arbuscular mycorrhiza fungal communities in the soil under Mediterranean conditions. Soil Tillage Res. 121, 63–67.

Brito, I., Carvalho, M., Alho, L., Caseiro, M., Goss, M.J., 2013a. Practical exploitation of mycorrhizal fungi in agricultural systems. Rethinking Agricultural Systems in the UK. Asp. Appl. Biol. 121, 25–30.

Brito, I., Carvalho, M., Goss, M.J., 2013b. Soil and weed management for enhancing arbuscular mycorrhiza colonisation of wheat. Soil Use Manage. 29, 540–546.

Brito, I., Carvalho, M., Alho, L., Goss, M.J., 2014. Managing arbuscular mycorrhizal fungi for bioprotection: Mn toxicity. Soil Biol. Biochem. 68, 78–84.

Brown, M.S., Bethlenfalvay, G.J., 1987. The *Glycine-Glomus-Rhizobium* symbiosis. VI. Photosynthesis in nodulated, mycorrhizal, or N- and P- fertilized soybean plants. Plant Physiol. 85, 120–123.

Brown, M.S., Bethlenfalvay, G.J., 1988. The *Glycine-Glomus-Rhizobium* symbiosis. VII. Photosynthetic nutrient-use efficiency in nodulated, mycorrhizal soybeans. Plant Physiol. 86, 1292–1297.

Bruinsma, J. (Ed.), 2003. World Agriculture: Towards 2015/2030. An FAO Perspective. Earthscan Publications Ltd., London. Available from: http://www.fao.org/docrep/005/y4252e/y4252e00.htm (accessed 08.12.15).

Bruinsma, J., 2011. The resources outlook: By how much do land, water and crop yields need to increase by 2050? In: Conforti, P. (Ed.), Looking Ahead in World Food and Agriculture: Perspectives to 2050. Food and Agriculture Organization of the United Nations, Rome, Italy, pp. 233–278. Available at: <http://www.fao.org/docrep/014/i2280e/i2280e.pdf>.

Brundrett, M.C., 2009. Mycorrhizal associations and other means of nutrition of vascular plants: understanding the global diversity of host plants by resolving conflicting information and developing reliable means of diagnosis. Plant Soil 320, 37–77.

Brundrett, M.C., Abbott, L.K., Jasper, D.A., 1999. Glomalean fungi from tropical Australia. I. Comparison of the effectiveness of isolation procedures. Mycorrhiza 8, 305–314.

Bucher, M., 2007. Functional biology of plant phosphate uptake at root and mycorrhiza interfaces. New Phytol. 173, 11–26.

Buee, M., Rossignol, M., Jauneau, A., Ranjeva, R., Bécard, G., 2000. The pre-symbiotic growth of arbuscular mycorrhizal fungi is induced by a branching factor partially purified from plant root exudates. Mol. Plant-Microbe Interact. J. 13, 693–698.

Bumb, B.L., Baanante, C.A., 1996. World trends in fertilizer use and projections to 2020. International Food Policy Research Institute, 2020 Brief 38, 4 pp.

Burleigh, S.H., Cavagnaro, T., Jakobsen, I., 2002. Functional diversity of arbuscular mycorrhizas extends to the expression of plant genes involved in P nutrition. J. Exp. Bot. 53, 1593–1601.

Cage, D.J., 2004. Infection and invasion of roots by symbiotic, nitrogen-fixing rhizobia during nodulation of temperate legumes. Microbiol. Mol. Biol. Rev. 68, 280–300.

Calado, J.M.G., Basch, G., Carvalho, M., 2010. Weed management in no-till winter wheat (*Triticum aestivum* L.). Crop Prot. 29, 1–6.

Calado, J.M.G., Basch, G., Barros, J.F.C., Carvalho, M., 2013. Weed management in winter wheat (*Triticum aestivum* L.) influenced by different soil tillage systems. Afr. J. Agric. Res. 8, 2551–2558.

Callaham, D.A., Torrey, J.G., 1981. The structural basis for infection of root hairs of *Trifolium repens* by Rhizobium. Can. J. Bot. 59, 1647–1664.

Calvo-Polanco, M., Sánchez-Romera, B., Aroca, R., Asins, M.J., Declerck, S., Dodd, I.C., Martínez-Andújar, C., Albacete, A., Ruiz-Lozano, J.M., 2016. Exploring the use of recombinant inbred lines in combination with beneficial microbial inoculants (AM fungus and PGPR) to improve drought stress tolerance in tomato. Environ. Exp. Bot. 131, 47–57.

Cameron, D.D., Neal, A.L., van Wees, S.C.M., Ton, J., 2013. Mycorrhiza–induced resistance: more than the sum of its parts? Trends Plant Sci. 18, 539–545.

Campanelli, A., Ruta, C., Mastro, G.D., Morone-Fortunato, I., 2013. The role of arbuscular mycorrhizal fungi in alleviating salt stress in *Medicago sativa* L. var. icon. Symbiosis 59, 65–76.

Campos, M.A.D., da Silva, F.S.B., Yano-Melo, A.M., de Melo, N.F., Pedrosa, E.M.R., Maia, L.C., 2013. Responses of guava plants to inoculation with arbuscular mycorrhizal fungi in soil infested with *Meloidogyne enterolobii*. Plant Pathol. J. 29, 242–248.

Cappellazzo, G., Lanfranco, L., Fitz, M., Wipf, D., Bonfante, P., 2008. Characterization of an amino acid permease from the endomycorrhizal fungus *Glomus mosseae*. Plant Physiol. 147, 429–437. Available from: http://dx.doi.org/10.1104/pp.108.117820.

Caravaca, F., Alguacil, M.M., Azcon, R., Roldan, A., 2006. Formation of stable aggregates in rhizosphere soil of *Juniperus oxycedrus*: effect of AM fungi and organic amendments. Appl. Soil Ecol. 33, 30–38.

Cardinale, B.J., Matulich, K.L., Hooper, D.U., Byrnes, J.E., Duffy, E., Gamfeldt, L., et al., 2011. The functional role of producer diversity in ecosystems. Am. J. Bot. 98, 572–592.

Cardoso, I.M., Kuyper, T.W., 2006. Mycorrhizas and tropical soil fertility. Agric. Ecosyst. Environ. 116, 72–84.

Carlsen, S.C.K., Understrup, A., Fomsgaard, I.S., Mortensen, A.G., Ravnskov, S., 2008. Flavonoids in roots of white clover: interaction of arbuscular mycorrhizal fungi and a pathogenic fungus. Plant Soil 302, 33–43.

Carvalho, M., 2013. The role of no-till and crop residues on sustainable arable crops production in southern portugal. In: Evelpidou N., Cordier St., Merino A., Figueiredo T., Centeri C. (Eds.), Runoff Erosion, 312–326. E.U Education and Culture DG, Lifelong learning programme.

Carvalho, M., Lourenço, E., 2014. Conservation agriculture—a portuguese case study. Review article. J. Agron. Crop Sci. 317–324.

Carvalho, M., Brito, I., Alho, L., Goss, M.J., 2015. Assessing the progress of colonization by arbuscular mycorrhiza of four plant species under different temperature regimes. J. Plant Nutr. Soil Sci. 178, 515–522.

Castillo, C.G., Rubio, R., Rouanet, J.L., Borie, F., 2006a. Early effects of tillage and crop rotation on arbuscular mycorrhizal fungal propagules in an Ultisol. Biol. Fertil. Soils 43, 83–92.

Castillo, P., Nico, A.I., Azcon-Aguilar, C., Del Rio Rincon, C., Calvet, C., Jimenez-Diaz, R.M., 2006b. Protection of olive planting stocks against parasitism of root-knot nematodes by arbuscular mycorrhizal fungi. Plant Pathol. 55, 705–713.

Cavagnaro, T.R., 2014. Impacts of compost application on the formation and functioning of arbuscular mycorrhizas. Soil Biol. Biochem. 78, 38–44.

Cavagnaro, T.R., Martin, A.W., 2011. Arbuscular mycorrhizas in southeastern Australian processing tomato farm soils. Plant Soil 340, 327–336.

Cavagnaro, T.R., Bender, S.F., Asghar, H.R., van der Heijden, M.G.A., 2015. The role of arbuscular mycorrhizas in reducing soil nutrient loss. Trends Plant Sci. 20, 283–290.

Celik, I., Barut, Z.B., Ortas, I., Gok, M., Demirbas, A., Tulun, Y., et al., 2011. Impacts of different tillage practices on some soil microbiological properties and crop yield under semi-arid Mediterranean conditions. Int. J. Plant Prod. 5, 237–254.

Chagnon, P.L., Bradley, R.L., Maherali, H., Klironomos, J.N., 2013. A trait-based framework to understand life history of mycorrhizal fungi. Trends Plant Sci. 18, 484–491.

Chaparro, J.M., Badri, D.V., Bakker, M.G., Sugiyama, A., Manter, D.K., Vivanco, J.M., 2013. Root exudation of phytochemicals in arabidopsis follows specific patterns that are developmentally programmed and correlate with soil microbial functions. PLoS ONE 8 (2), e55731.

Charest, C., Dalpe, Y., Brown, A., 1993. The effect of vesicular-arbuscular mycorrhizae and chilling on 2 hybrids of *Zea mays* L. Mycorrhiza 4, 89–92.

Charreu, C., 1972. Problemes poses par l'utilization agricole des sols tropicaux par des cultures annuelles. Agron. Trop. 27, 905–929.

Chaudhary, V.B., Lau, M.K., Johnson, N.C., 2008. Macroecology of microbes—biogeography of the Glomeromycota. In: Varma, A. (Ed.), Mycorrhiza. Springer, Berlin, pp. 529–562.

Chauhan, B.S., Gill, G., Preston, C., 2006. Seedling recruitment pattern and depth of recruitment of 10 weed species in minimum tillage and no-till seeding systems. Weed Sci. 54, 658–668.

Christie, P., Kilpatrick, D.J., 1992. Vesicular-arbuscular mycorrhiza infection in cut grassland following long-term slurry application. Soil Biol. Biochem. 24, 325–330.

Cicatelli, A., Torrigiani, P., Todeschini, V., Biondi, S., Castiglione, S., Lingua, G., 2014. Arbuscular mycorrhizal fungi as a tool to ameliorate the phytoremediation potential of poplar: biochemical and molecular aspects. Iforest-Biogeosciences and Forestry 7, 333–341.

Clapp, J.P., Rodriguez, A., Dodd, J.C., 2001. Inter- and intra-isolate rRNA large subunit variation in *Glomus coronatum* spores. New Phytol. 149, 539–554.

Clapperton, M.J., Reid, D.M., Parkinson, D., 1990. Effects of sulfur-dioxide fumigation on *Phleum pratense* and vesicular-arbuscular mycorrhizal fungi. New Phytol. 115, 465–469.

Clark, R.B., Zeto, S.K., 2000. Mineral acquisition by arbuscular mycorrhizal plants. J. Plant. Nutr. 23, 867–902.

Clark, R.B., Zobel, R.W., Zeto, S.K., 1999. Effects of mycorrhizal fungus isolates on mineral acquisition by Panicum virgatum in acidic soil. Mycorrhiza 19, 167–176.

Clay, K., 2014. Defensive symbiosis: a microbial perspective. Funct. Ecol. 28, 293–298.

Clements, D.R., Benott, D.L., Murphy, S.D., Swanton, C.J., 1996. Tillage effects on weed seed return and seedbank composition. Weed Sci. 44, 314–322.

Cluett, H.C., Boucher, D.H., 1983. Indirect mutualism in the legume-*Rhizobium* mycorrhizal fungus interaction. Oecologia (Berlin) 59, 405–408.

Compant, S., van der Heijden, M.G.A., Sessitsch, A., 2010. Climate change effects on beneficial plant-microorganism interactions. FEMS Microbiol. Ecol. 73, 197–214.

Cooper, J.E., 2007. Early interactions between legumes and rhizobia: disclosing complexity in a molecular dialogue. J. Appl. Microbiol. 103, 1355–1365.

Cordell, D., Drangert, J.O., White, S., 2009. The story of phosphorus: global food security and food for thought. Glob. Environ. Change 19, 292–305.

Cordoba, A., De Mendonça, M., Stürmer, S., Rygiewicz, P., 2001. Diversity of arbuscular mycorrhizal fungi along a sand dune stabilization gradient: a case study at Praia da Joaquina, Ilha de Santa Catarina, South Brazil. Mycoscience 42, 379–387.

Corkidi, L., Rowland, D.L., Johnson, N.C., Allen, E.B., 2002. Nitrogen fertilization alters the functioning of arbuscular mycorrhizas at two semiarid grasslands. Plant Soil 240, 299–310.

Cornejo, P., Perez-Tienda, J., Meier, S., Valderas, A., Borie, F., Azcon-Aguilar, C., et al., 2013. Copper compartmentalization in spores as a survival strategy of arbuscular mycorrhizal fungi in Cu-polluted environments. Soil Biol. Biochem. 57, 925–928.

Corradi, N., Bonfante, P., 2012. The arbuscular mycorrhizal symbiosis: origin and evolution of a beneficial plant infection. PLoS Pathog. 8 (4), e1002600.

Creamer, R.E., Hannula, S.E., Van Leeuwen, J.P., Stone, D., Rutgers, M., Schmelz, R.M., et al., 2016. Ecological network analysis reveals the inter-connection between soil biodiversity and ecosystem function as affected by land use across Europe. Appl. Soil Ecol. 97, 112–124. Available from: http://dx.doi.org/10.1016/j.apsoil.2015.08.006.

Croll, D., Wille, L., Gamper, H.A., Mathimaran, N., Lammers, P.J., Corradi, N., et al., 2008. Genetic diversity and host plant preferences revealed by simple sequence repeat and mitochondrial markers in a population of the arbuscular mycorrhizal fungus *Glomus intraradices*. New Phytol. 178, 672–687.

Cruz, A.F., Ishii, T., 2011. Arbuscular mycorrhizal fungal spores host bacteria that affect nutrient dynamics and biocontrol of soil-borne plant pathogens. Biol. Open 1, 52–57.

Cruz, A.F., Horii, S., Ochiai, S., Yasuda, A., Ishii, T., 2008. Isolation and analysis of bacteria associated with spores of *Gigaspora margarita*. J. Appl. Microbiol. 104, 1711–1717.

Curaqueo, G., Acevedo, E., Cornejo, P., Seguel, A., Rubio, R., Borie, F., 2010. Tillage effect on soil organic matter, mycorrhizal hyphae and aggregates in a mediterranean agroecosystem. Revista de la ciencia del suelo y nutrición vegetal 10, 12–21.

Daniell, T.J., Husband, R., Fitter, A.H., Young, J.P.W., 2001. Molecular diversity of arbuscular mycorrhizal fungi colonising arable crops. FEMS Microbiol. Ecol. 36, 203–209.

Darbyshire, J.F., Greaves, M.P., 1967. Protozoa and bacteria in the rhizosphere of *Sinapis alba* L., *Trifolium repens* L., and *Lolium perenne* L. Can. J. Microbiol. 13, 1057–1068.

David–Schwartz, R., Badani, H., Smadar, W., Levy, A.A., Galili, G., Kapunik, Y., 2001. Identification of a novel genetically controlled step in mycorrhizal colonization: plant resistance to infection by fungal spores but not extra–radical hyphae. Plant J. 27, 561–569.

Dazzo, F.B., Yanke, W.E., Brill, W.J., 1978. Trifolin: a rhizobium recognition protein from white clover. Biochim. Biophys. Acta 539, 276–286.

de Boer, R.F., Steed, G.R., Kollmorgen, J.F., Macauley, B.J., 1993. Effects of rotation, stubble retention and cultivation on take-all and eyespot of wheat in northeastern Victoria, Australia. Soil Tillage Res. 25, 263–280.

de Boer, W., Folman, L.B., Summerbell, R.C., Boddy, L., 2005. Living in a fungal world: impact of fungi on soil bacterial niche development. FEMS Microbiol. Rev. 29, 795–811.

de Boer, W., Hundscheid, M.P.J., Paulien, J.A., Klein Gunnewiek, P.J.A., de Ridder-Duine, A.S., Thion, C., et al., 2015. Antifungal rhizosphere bacteria can increase as response to the presence of saprotrophic fungi. PLoS ONE 10 (9), e0137988. Available from: http://dx.doi.org/10.1371/journal.pone.0137988.

DeHaan, L.R., Russelle, M.P., Sheaffer, C.C., Ehlke, N.J., 2002. Kura clover and birdsfoot trefoil response to soil pH. Commun. Soil Sci. Plant Anal. 33, 1435–1449.

De Jong, E., Douglas, J.T., Goss, M.J., 1983. Gaseous diffusion in shrinking soils. Soil Sci. 136, 10–18.

De Ley, J., De Smedt, J., 1978. Improvements of the membrane filter method for DNA:rRNA hybridization. Antonie. Van. Leeuwenhoek. 41, 287–307. Available from: http://dx.doi.org/10.1007/BF02565064.

Desirò, A., Salvioli, A., Ngonkeu, E.L., Mondo, S.J., Epis, S., Faccio, A., et al., 2014. Detection of a novel intracellular microbiome hosted in arbuscular mycorrhizal fungi. ISME J. 8, 257–270.

de Varennes, A., Goss, M.J., 2007. The tripartite symbiosis between legumes, rhizobia and indigenous mycorrhizal fungi is more efficient in undisturbed soil. Soil Biol. Biochem. 39, 2603–2607.

de Varennes, A., Carneiro, J.P., Goss, M.J., 2001. Characterization of manganese toxicity in two species of annual medics. J. Plant Nutr. 24, 1947–1955.

de Vrieze, J., 2015. The littlest farmhands. Science 349, 680–683.

Dias, T., Dukesa, A., Antunes, P.M., 2015. Accounting for soil biotic effects on soil health and crop productivity in the design of crop rotations. J. Sci. Food. Agric. 95, 447–454.

Diedhiou, P.M., Hallmann, J., Oerke, E.C., Dehne, H.W., 2003. Effect of arbuscular mycorrhizal fungi and a non-pathogenic *Fusarium oxysporum* on *Meloidogyne incognita* infestation of tomato. Mycorrhiza 13, 199–204.

Dixon, J., Gulliver, A., Gibbon, D., 2001. Farming Systems and Poverty: Improving Farmers' Livelihoods in a Changing World. FAO &. World Bank, Rome, Italy & Washington, DC.

Dobereiner, J., 1966. Manganese toxicity effects on nodulation and nitrogen fixation of beans (*Phaseolus vulgaris* L.), in acid soils. Plant Soil 24, 153–166.

Dodd, J.C., 2000. The role of arbuscular mycorrhizal fungi in agro- and natural ecosystems. Outlook. Agric. 29, 55–62.

Dodd, J.C., Boddington, C.L., Rodriguez, A., Gonzalez–Chavez, C., Mansur, I., 2000. Mycelium of arbuscular mycorrhizal fungi (AMF) from different genera: form, function and detection. Plant Soil 226, 131–151.

Doolittle, W.F., Woese, C.R., Sogin, M.L., Bonen, L., Stahl, D., 1975. Sequence studies on 16S ribosomal RNA from a blue-green alga. J. Mol. Evol. 4, 307–315. Available from: http://dx.doi.org/10.1007/BF01732533.

Douds, D.D., Galvez, L., Franke-Snyder, M., Reider, C., Drinkwater, L.E., 1997. Effect of compost addition and crop rotation point upon VAM fungi. Agric. Ecosyst. Environ. 65, 257–266.

Douhan, G.W., Petersen, C., Bledsoe, C.S., Rizzo, D.M., 2005. Contrasting root associated fungi of three common oak-woodland plant species based on molecular identification: host specificity or non-specific amplification? Mycorrhiza 15, 365–372. Available from: http://dx.doi.org/10.1007/s00572-004-0341-2.

Downie, J.A., 2010. The rolesof extracellular proteins, polysaccharides and signals in the interactions of rhizobia with legume roots. FEMS Microbiol. Rev. 34, 150–170.

Drew, E.A., Murray, R.S., Smith, S.E., Jakobsen, I., 2003. Beyond the rhizosphere: growth and function of arbuscular mycorrhizal external hyphae in sands of varying pore sizes. Plant Soil 251, 105–114.

Drew, E.A., Murray, R.S., Smith, S.E., 2006. Functional diversity of external hyphae of AM fungi: ability to colonise new hosts is influenced by fungal species, distance and soil conditions. Appl. Soil Ecol. 32, 350–365.

Driver, J.D., Holben, W.E., Rillig, M.C., 2005. Characterization of glomalin as a hyphal wall component of arbuscular mycorrhizal fungi. Soil Biol. Biochem. 37, 101–106.

Druille, M., Cabello, M.N., Omacini, M., Golluscio, R.A., 2013. Arbuscular mycorrhizal fungi are directly and indirectly affected by glyphosate application. Appl. Soil Ecol. 72, 143–149.

Druille, M., Cabello, M.N., Omacini, M., Golluscio, R.A., 2015. Glyphosate vulnerability explains changes in root-symbionts propagules viability in pampean grasslands. Agric. Ecosyst. Environ. 202, 48–55.

Duc, G., Trouvelot, A., Gianinazzi-Pearson, V., Gianinazzi, S., 1989. First report of non-mycorrhizal plants mutants ($Myc^-$) obtained in (*Pisum sativa* L.) and Fababean (*Vicia faba* L.). Plant Sci. 60, 215–222.

Dumanski, J., 2015. Evolving concepts and opportunities in soil conservation. Int. Soil Water Conserv. Res. 3, 1–14.

Dybvig, K., Voelker, L.L., 1996. Molecular biology of mycoplasmas. Ann. Rev. Microbiol. 50, 25–57.

Eason, W.R., Scullion, J., Scott, E.P., 1999. Soil parameters and plant responses associated with arbuscular mycorrhizas from contrasting grassland management regimes. Agric. Ecosyst. Environ. 73, 245–255.

Edwards, W.M., Shipitalo, M.J., Traina, S.J., Edwards, C.A., Owen, L.B., 1992. Role of *Lumbrlcus terrestris* (L.) burrows on quality of infiltrating water. Soil Biol. Biochem. 24, 1555–1561.

Egerton-Warburton, L., Allen, E., 2000. Shifts in arbuscular mycorrhizal communities along ananthropogenic nitrogen deposition gradient. Ecol. Appl. 10, 484–496.

Egerton-Warburton, L.M., Johnson, N.C., Allen, E.B., 2007. Mycorrhizal community dynamics following nitrogen fertilization: a cross-site test in five grasslands. Ecol. Monogr. 77, 527–544.

Ehinger, Koch, A.M., Sanders, I.R., 2009. Changes in arbuscular mycorrhizal fungal phenotypes and genotypes in response to plant species identity and phosphorus concentration. New Phytol. 184, 412–423.

Ehlers, W., Goss, M., 2016. Water Dynamics in Plant Production, second ed. CABI Publishing, Wallingford.

Eisenhauer, N., König, S., Sabais, A.C.W., Renker, C., Buscot, F., Scheu, S., 2009. Impacts of earthworms and arbuscular mycorrhizal fungi (*Glomus intraradices*) on plant performance are not interrelated. Soil Biol. Biochem. 41, 167–176.

Elbon, A., Whalen, J.K., 2015. Phosphorus supply to vegetable crops from arbuscular mycorrhizal fungi: a review. Biol. Agric. Hortic. 31, 73–90.

El-Hassanin, A.S., Lynd, J.Q., 1985. Soil fertility effects with tripartite symbiosis for growth, nodulation and nitrogenase activity of Vicia faba L. J. Plant Nutr. 8, 491–504.

Ellouze, W., Hamel, C., Vujanovic, V., Gan, Y., Bouzid, S., St-Arnaud, M., 2013. Chickpea genotypes shape the soil microbiome and affect the establishment of the subsequent durum wheat crop in the semiarid North American Great Plains. Soil Biol. Biochem. 63, 129–141.

Eriksson, J., Håkansson, I., Danfors, B., 1974. The effect of soil compaction on soil structure and crop yields. Bulletin 354 Swedish Institute of Agricultural Engineering, Uppsala.

Estaún, V., Calvet, C., Hayman, D.S., 1987. Influence of plant genotype on mycorrhizal infection—response of 3 pea cultivars. Plant Soil 103, 295–298.

Eswaran, H., Lal, R., Reich, P.F., 2001. Land degradation: an overview. In: Bridges, E.M., Hannam, I.D., Oldeman, L.R., Pening de Vries, F.W.T., Scherr, S.J., Sompatpanit, S. (Eds.), Responses to Land Degradation. Proceedings of Second International Conference on Land Degradation and Desertification, Khon Kaen, Thailand. Oxford Press, New Delhi, India.

Evans, D.G., Miller, M.H., 1990. The role of the external mycelial network in the effect of soil disturbance upon vesicular-arbuscular mycorrhizal colonization of maize. New Phytol. 114, 65–71.

Evans, J., Scott, B.J., Lill, W.J., 1987. Manganese tolerance in subterranean clover (*Trifolium subterraneum* L.) genotype grown with ammonium nitrate or symbiotic nitrogen. Plant Soil 97, 207–215.

Evelin, H., Kapoor, R., Giri, B., 2009. Arbuscular mycorrhizal fungi in alleviation of salt stress: a review. Ann. Bot. (Lond). 104, 1263–1280.

Fairchild, G.L., Miller, M.H., 1988. Vesicular-arbuscular mycorrhizas and the soil-disturbance induced reduction of nutrient absorption in maize II. Development of the effect. New Phytol. 110, 75–84.

Fairchild, G.L., Miller, M.H., 1990. Vesicular–arbuscular mycorrhizas and the soil–disturbance–induced reduction of nutrient absorption in maize. New Phytol. 114, 641–650.

FAO (Food and Agriculture Organization of the United Nations), 2009. How to Feed the World in 2050. Available from: <http://www.fao.org/fileadmin/templates/wsfs/docs/expert_paper/How_to_Feed_the_World_in_2050.pdf> (accessed 04.12. 15.).

FAO (Food and Agriculture Organization of the United Nations), 2011. The State of the World's Land and Water Resources for Food and Agriculture (SOLAW)—managing systems at risk; Earthscan, Abingdon, UK. Available from: <http://www.fao.org/docrep/017/i1688e/i1688e00.htm>.

FAO (Food and Agriculture Organization of the United Nations), 2015a. State of Food Insecurity in the World FAO, Rome, 2015. Available from: <http://www.fao.org/hunger/key-messages/en/> (accessed 04.12.15.).

FAO (Food and Agriculture Organization of the United Nations), 2015b. Agriculture and Consumer Protection Department. Conservation Agriculture. <http://www.fao.org/ag/ca/1a.html>.

FAO (Food and Agriculture Organization of the United Nations), and ITPS (Intergovernmental Technical Panel on Soils), 2015. Status of the World's Soil Resources (SWSR)—Main Report. Food and Agriculture Organization of the United Nations and Intergovernmental Technical Panel on Soils, Rome, Italy.

FAOSTAT., 2015. FAOSTAT database. Available at: < http://www.fao.org/faostat/en/#home> (accessed 03.12.15.).

Farmer, M.J., Li, X., Feng, G., Zhao, B., Chatagnier, O., Gianinazzi, S., et al., 2007. Molecular monitoring of field-inoculated AMF to evaluate persistence in sweet potato crops in China. Appl. Soil Ecol. 35, 599–609. Available from: http://dx.doi.org/10.1016/j.apsoil.2006.09.012.

Fernández, I., Merlos, M., López-Ráez, J.A., Martinez-Medina, A., Ferrol, N., Azcón, C., et al., 2014. Defense related phytohormones regulation in arbuscular mycorrhiza symbiosis depends on the partner genotypes. J. Chem. Ecol. 40, 791–803.

Fierer, N., Schimela, J.P., Holden, P.A., 2003. Variations in microbial community composition through two soil depth profiles. Soil Biol. Biochem. 35, 167–176.

Filion, M., St-Arnaud, M., Fortin, J.A., 1999. Direct interaction between the arbuscular mycorrhizal fungus *Glomus intraradices* and different rhizosphere microorganisms. New Phytol. 141, 525–533.

Filion, M., St-Arnaud, M., Jabaji-Hare, S.H., 2003. Quantification of *Fusarium solani* f. sp. phaseoli in mycorrhizal bean plants and surrounding mycorrhizosphere soil using real-time polymerase chain reaction and direct isolations on selective media. Phytopathology 93, 229–235.

Fitter, A.H., 1988. Water relations of red clover Trifolium pratense L. as affected by VA mycorrhizal infection and phosphorus supply before and during drought. J. Exp. Bot. 39, 595–603.

Fitter, A.H., 2005. Darkness visible: reflections on underground ecology. J. Ecol. 93, 231–243.

Fitter, A.H., Garbaye, J., 1994. Interactions between mycorrhizal fungi and other soil organisms. Plant Soil 159, 123–132.

Fitter, A.H., Merryweather, J.W., 1992. Why are some plants more mycorrhizal than others? An ecological enquiry. In: Read, D.J., Lewis, D.H., Fitter, A.H., Alexander, I.J. (Eds.), Mycorrhizas in Ecosystems. CAB International, Wallingford, pp. 26–36.

Foley, J.A., DeFries, R., Asner, G.P., Barford, C., Bonan, G., Carpenter, S.R., et al., 2005. Global consequences of land use. Science 309, 570–574.

Foley, J.A., Ramankutty, N., Brauman, K.A., Cassidy, E.S., Gerber, J.S., Johnston, M., et al., 2011. Solutions for a cultivated planet. Nature 478, 337–342.

Foo, E., Ross, J.J., Jones, W.T., Reid, J.B., 2013. Plant hormones in arbuscular mycorrhizal symbioses: an emerging role for gibberellins. Ann. Bot. (Lond). 111, 769–779.

Frey, B., Schüepp, H., 1992. Transfer of symbiotically fixed nitrogen from berseem (*Trifolium alexandrium* L.) to maize via vesicular-arbuscular mycorrhizal hyphae. New Phytol. 122, 447–454.

Frey-Klett, P., Garbaye, J., Tarkka, M., 2007. The mycorrhiza helper bacteria revisited. New Phytol. 176, 22–36.

Friese, C.F., Allen, M.F., 1991. The spread of VA mycorrhizal fungal hyphae in the soil: inoculum types and external hyphal architecture. Mycologia 83, 409–418.

Froud-Williams, R.J., Chancellor, R.J., Drennan, D.S.H., 1981. Potential changes in weed floras associated with reduced-cultivation systems for cereal production in temperate regions. Weed Res. 21, 99–109.

Froud-Williams, R.J., Drennan, D.S.H., Chancellor, R.J., 1983. Influence of cultivation regime on weed floras of arable cropping systems. J. Appl. Ecol. 20, 187–197.

Gai, J., Gao, W., Liu, L., Chen, Q., Feng, G., Zhang, J., et al., 2015. Infectivity and community composition of arbuscular mycorrhizal fungi from different soil depths in intensively managed agricultural ecosystems. J. Soils Sediments 15, 1200–1211.

Galván, G.A., Parádi, I., Burger, K., Baar, J., Kuyper, T.W., Scholten, O.E., et al., 2009. Molecular diversity of arbuscular mycorrhizal fungi in onion roots from organic and conventional farming systems in the Netherlands. Mycorrhiza 19, 317–328. Available from: http://dx.doi.org/10.1007/s00572-009-0237-2.

Galvez, L., Douds Jr, D.D., Drinkwater, L.E., Wagoner, P., 2001. Effect of tillage and farming system upon VAM fungus populations and mycorrhizas and nutrient uptake of maize. Plant Soil 228, 299–308.

Gamalero, E., Berta, G., Massa, N., Glick, B.R., Lingua, G., 2008. Synergistic interactions between the ACC deaminase-producing bacterium *Pseudomonas putida* UW4 and the AM fungus *Gigaspora rosea* positively affect cucumber plant growth. FEMS Microbiol. Ecol. 64, 459–467.

Gamper, H., Leuchtmann, A., 2007. Taxon-specific PCR primers to detect two inconspicuous arbuscular mycorrhizal fungi from temperate agricultural grassland. Mycorrhiza 17, 145–152. Available from: http://dx.doi.org/10.1007/s00572-006-0092-3.

Gamper, H.A., van der Heijden, M.G.A., Kowalchuk, G.A., 2010. Molecular trait indicators: moving beyond phylogeny in arbuscular mycorrhizal ecology. New Phytol. 185, 67–82. Available from: http://dx.doi.org/10.1111/j.1469-8137.2009.03058.x.

Gamper, H.A., Walker, C., Schüßler, A., 2009. Diversispora celata sp. nov: molecular ecology and phylotaxonomy of an inconspicuous arbuscular mycorrhizal fungus. New Phytol. 182, 495–506. Available from: http://dx.doi.org/10.1111/j.1469-8137.2008.02750.x.

Gamper, H.A., Young, J.P.W., Jones, D.L., Hodge, A., 2008. Real-time PCR and microscopy: Are the two methods measuring the same unit of arbuscular mycorrhizal fungal abundance? Fungal Genet. Biol. 45, 581–596. Available from: http://dx.doi.org/10.1016/j.fgb.2007.09.007.

Gange, A., 2000. Arbuscular mycorrhizal fungi, Collembola and plant growth. Tree 15, 369–372.

Gans, J., Wolinsky, M., Dunbar, J., 2005. Computational improvements reveal great bacterial diversity and high metal toxicity in soil. Science 309, 1387–1390.

Gao, X., Kuyper, T.W., Zou, C., Zhang, F., Hoffland, E., 2007. Mycorrhizal responsiveness of aerobic rice genotypes is negatively correlated with their zinc uptake when nonmycorrhizal. Plant Soil 290, 283–291.

Garbaye, J., 1994. Helper bacteria: a new dimension to the mycorrhizal symbiosis. New Phytol. 128, 197–210.

Garg, N., Chandel, S., 2010. Arbuscular mycorrhizal networks: process and functions. A review. Agron. Sustain. Dev. 30, 581–599.

Gaskin, J.W., Speir, R.A., Harris, K., Das, K.C., Lee, R.D., Morris, L.A., et al., 2010. Effect of peanut hull and pine chip biochar on soil nutrients, corn nutrient status, and yield. Agron. J. 102, 623–633.

Gavito, M.E., Miller, M.H., 1998. Changes in mycorrhiza development in maize induced by crop management practices. Plant Soil 198, 185–192.

Genre, A., Chabaud, M., Timmers, T., Bonfante, P., Barker, D.G., 2005. Arbuscular mycorrhizal fungi elicit a novel intracellular apparatus in *Medicago truncatula* root epidermal cells before infection. Plant Cell 17, 3489–3499.

Genre, A., Chabaud, M., Faccio, A., Barker, D.G., Bonfante, P., 2008. Prepenetration apparatus assembly precedes and predicts the colonization patterns of arbuscular mycorrhizal fungi within the root cortex of both Medicago truncatula and *Daucus carota*. Plant Cell 20, 1407–1420.

George, E., 2000. Nutrient uptake: contribution of arbuscular mycorrhizal fungi to plant mineral nutrition. In: Kapulnik, Y., Douds Jr., D.D. (Eds.), Arbuscular Mycorrhizas: Physiology and Function. Kluwer Academic Publishers, Dordrecht, pp. 307–343.

George, E., Haussler, K.-U., Vetterlein, D., Gorgus, E., Marschiner, H., 1992. Water and nutrient translocation by hyphae of *Glomus mosseae*. Can. J. Bot. 70, 12–30.

George, E., Marschner, H., Jakobsen, I., 1995. Role of arbuscular mycorrhizal fungi in uptake of phosphorus and nitrogen from soil. Crit. Rev. Biotechnol. 15, 257–270.

Ghorbani-Nasrabadi, R., Greiner, R., Alikhani, H.A., Hamedi, J., Yakhchali, B., 2013. Distribution of actinomycetes in different soil ecosystems and effect of media composition on extracellular phosphatase activity. J. Soil Sci. Plant Nutr. 13, 223–236.

Gianinazzi, S., Gollotte, A., Binet, M.N., van Tuinen, D., Redecker, D., Wipf, D., 2010. Agroecology: the key role of arbuscular mycorrhizas in ecosystem services. Mycorrhiza 20, 519–530. Available from: http://dx.doi.org/10.1007/s00572-010-0333-3.

Gianinazzi-Pearson, V., Dénarié, J., 1997. Red carpet genetic programmes for root endosymbioses. Trends Plant Sci. 2, 371–372.

Gianinazzi-Pearson, V., Gianinazzi, S., Guillemin, J.P., Trouvelot, A., Duc, G., 1991. Genetic and cellular analysis of resistance to vesicular-arbuscular (VA) mycorrhizal fungi in pea mutants. In: Hennecke, H., Verma, D.P.S. (Eds.), Advances in Molecular Genetics of Plant-Microbe Interactions. Kluwer Academic Publishers, London, pp. 336–342.

Gianinazzi-Pearson, V., Gollotte, A., Cordier, C., Gianinazzi, S., 1996. Root defense responses in relation to cell and tissue invasion by symbiotic microorganisms: cytological investigations. In: Nicoie, M., Gianinazzi-Pearson, V. (Eds.), Histology, Ultrastructure and Molecular Cytology of Plant-Microorganism Interactions. Kluwer Academic Publishers, Dordrecht, pp. 127–191.

Gill, W.R., 1971. Economic Assessment of Soil Compaction. St. Joseph: ASAE Monograph.

Giovannetti, M., Gianinazzi-Pearson, V., 1994. Biodiversity in arbuscular mycorrhizal fungi. Mycol. Res. 98, 705–715.

Giovannetti, M., Azzolini, D., Citernesi, A.S., 1999. Anastomosis formation and nuclear and protoplasmic exchange in arbuscular mycorrhizal fungi. Appl. Environ. Microbiol. 65, 5571–5575.

Giovannetti, M., Fortuna, P., Citernesi, A.S., Morini, S., Nuti, M.P., 2001. The occurrence of anastomosis formation and nuclear exchange in intact arbuscular mycorrhizal networks. New Phytol. 151, 717–724.

Giovannetti, M., Sbrana, C., Avio, L., Strani, P., 2004. Patterns of below–ground plant interconnections established by means of arbuscular mycorrhizal mycorrhizal networks. New Phytol. 164, 175–181.

Glenn, S.A., Gurich, N., Feeney, M.A., González, J.E., 2007. The ExpR/Sin quorum-sensing system controls succinoglycan production in Sinorhizobium meliloti. J. Bacteriol. 189, 7077–7088.

Gobbato, E., 2015. Recent developments in arbuscular mycorrhizal signaling. Curr. Opin. Plant. Biol. 26, 1–7.

Godfray, H.C.J., Beddington, J.R., Crute, I.R., Haddad, L., Lawrence, D., Muir, J.F., et al., 2010. Food security: the challenge of feeding 9 billion people. Science 327, 812–818.

Göhre, V., Paszkowski, U., 2006. Contribution of the arbuscular mycorrhizal symbiosis to heavy metal phytoremediation. Planta 223, 1115–1122.

Gollner, M., Friedel, J.K., Freyer, B., 2004. Auswirkungen acker- und pflanzenbaulicher Massnahmen auf die arbuskuläre Mykorrhiza im Ökologischen Landbau. Mitt. Ges. Pflanzenbauwiss. 16, 25–36.

Gollotte, A., van Tuinen, D., Atkinson, D., 2004. Diversity of arbuscular mycorrhizal fungi colonising roots of the grass species *Agrostis capillaris* and *Lolium perenne* in a field experiment. Mycorrhiza 14, 111–117. Available from: http://dx.doi.org/10.1007/s00572-003-0244-7.

Gormsen, D., Olsson, P.A., Hedlund, K., 2004. The influence of collembolans and earthworms on AM fungal mycelium. Appl. Soil Ecol. 27, 211–220.

Gosling, P., Ozaki, A., Jones, J., Turner, M., Rayns, F., Bending, G.D., 2010. Organic management of tilled agricultural soils results in a rapid increase in colonisation potential and spore populations of arbuscular mycorrhizal fungi. Agric. Ecosyst. Environ. 139, 273–279.

Goss, M.J., 1991. Consequences of the effects of roots on soil. In: Atkinson, D. (Ed.), Plant Root Growth. Blackwell Scientific Publications, Oxford, pp. 171–186.

Goss, M.J., de Varennes, A., 2002. Soil disturbance reduces the efficacy of mycorrhizal associations for early soybean growth and $N_2$ fixation. Soil Biol. Biochem. 34, 1167–1173.

Goss, M.J., Kay, B.D., 2005. Soil aggregation. In: Wright, S.F., Zobel, R.W. (Eds.), Roots and Soil Management: Interactions Between Root and the Soil. Agronomy Monograph No 48. ASA, CSSA, and SSSA, Madison, WI, pp. 163–180.

Goulding, K., Jarvis, S., Whitmore, A., 2008. Optimizing nutrient management for farm systems. Philos. Trans. R. Soc., B 363, 667–680.

Govindarajulu, M., Pfeffer, P.E., Jin, H.R., Abubaker, J., Douds, D.D., Allen, J.W., et al., 2005. Nitrogen transfer in the arbuscular mycorrhizal symbiosis. Nature 435, 819–823.

Graham, J.H., Linderman, R.G., Menge, J.A., 1982. Development of external hyphae by different isolates of mycorrhizal *Glomus spp.* in relation to root colonization and growth of troyer citrange. New Phytol. 91, 183–189.

Griffiths, B.S., Philippot, L., 2013. Insights into the resistance and resilience of the soil microbial community. FEMS Microbiol. Rev. 37, 112–129.

Griffiths, B.S., Ritz, K., 1988. A technique to extract, enumerate and measure protozoa from mineral soils. Soil Biol. Biochem. 20, 163–173.

Griffiths, B.S., Ritz, K., Bardgett, R.D., Cook, R., Christensen, S., Ekelund, F., et al., 2000. Ecosystem response of pasture soil communities to fumigation-induced microbial diversity reductions: an examination of the biodiversity–ecosystem function relationship. Oikos 90, 279–294.

Grime, J.P., 1979. Plant Strategies and Vegetative Processes. John Wiley & Sons, New York.

Guan, D., Stacey, N., Liu, C., Wen, J., Mysore, K.S., Torres-Jerez, I., et al., 2013. Rhizobial infection is associated with the development of peripheral vasculature in nodules of *Medicago truncatula*. Plant Physiol. 162, 107–115.

Guether, M., Neuhäuser, B., Balestrini, R., Dynowski, M., Ludewig, U., Bonfante, P., 2009. A mycorrhizal-specific ammonium transporter from *Lotus japonicus* acquires nitrogen released by arbuscular mycorrhizal fungi. Plant Physiol. 150, 73–83.

Guinel, F.C., Geil, R.D., 2002. A model for the development of the rhizobial and arbuscular mycorrhizal symbioses in legumes and its use to understand the roles of ethylene in the establishment of these two symbioses. Can. J. Bot. 80, 695–720.

Habte, M., 1989. Impact of simulated erosion on the abundance and activity of indigenous vesicular-arbuscular mycorrhizal endophytes in an oxisol. Biol. Fertil. Soils 7, 164–167.

Habte, M., Zhang, Y.C., Schmitt, D.P., 1999. Effectiveness of *Glomus* species in protecting white clover against nematode damage. Can. J. Bot.-Rev. Can. Bot. 77, 135–139.

Habteselassie, M.Y., Xu, L., Norton, J.M., 2013. Ammonia-oxidizer communities in an agricultural soil treated with contrasting nitrogen sources. Front. Microbiol. 4, 326. Available from: http://dx.doi.org/10.3389/fmicb.2013.00326.

Hajiboland, R., 2013. Role of arbuscular mycorrhiza in amelioration of salinity. In: Ahmad, P., Azooz, M.M., Prasad, M.N.V. (Eds.), Salt Stress in Plants: Signalling, Omics and Adaptations. Springer, New York, pp. 301–354.

Hajiboland, R., Aliasgharzad, N., Barzeg Har, R., 2009. Influence of arbuscular mycorrhizal fungi on uptake of Zn and P by two contrasting rice genotypes. Plant, Soil Environ. 55, 93–100.

Hallett, P.D., Feeney, D.S., Bengough, A.G., Rillig, M.C., Scrimgeour, C.M., Young, I.M., 2009. Disentangling the impact of AM fungi versus roots on soil structure and water transport. Plant Soil 314, 183–196.

Hamel, C., 1996. Prospects and problems pertaining to the management of arbuscular mycorrhizae in agriculture. Agric. Ecosyst. Environ. 60, 197–210.

Hamel, C., Dalpé, Y., Lapierre, C., Simard, R.R., Smith, D.L., 1994. Composition of the vesicular-arbuscular mycorrhizal fungi population in an old meadow as affected by pH, phosphorus and soil disturbance. Agric. Ecosyst. Environ. 49, 223–231.

Harinikumar, K.M., Bagyaraj, D.J., 1988. Effect of crop rotation on native vesicular arbuscular mycorrhizal propagules in soil. Plant Soil 110, 77–80.

Harrier, L.A., Watson, C.A., 2003. The role of arbscular mycorrhizal fungi in sustainable cropping systems. Adv. Agron. 79, 185–225.

Harrier, L.A., Watson, C.A., 2004. The potential role of arbuscular mycorrhizal (AM) fungi in the bioprotection of plants against soil-borne pathogens in organic and/or other sustainable farming systems. Pest. Manag. Sci. 60, 149–157.

Harris, D.R., Pacovsky, S., Paul, E.A., 1985. Carbon economy of soybean-*Rhizobium-Glomus* associations. New Phytol. 101, 427–440.

Harris, K.K., Boerner, R.E.J., 1990. Effects of belowground grazing by collembola on growth, mycorrhizal infection and P uptake of *Geranium robertianum*. Plant Soil 129, 203–210.

Harrison, M.J., 1999a. Molecular and cellular aspects of the arbuscular mycorrhizal symbiosis. Ann. Rev. Plant Physiol. Plant Mol. Biol. 50, 361–389.

Harrison, M.J., 1999b. Biotrophic interfaces and nutrient transport in plant/fungal symbioses. J. Exp. Bot. 50 (Special Issue), 1013–1022.

Harrison, M.J., 2005. Signaling in the arbuscular mycorrhizal symbiosis. Annu. Rev. Microbiol. 59, 19–42.

Harrison, M.J., Dixon, R.A., 1993. Isoflavonoid accumulation and expression of defence gene transcripts during the establishment of vesicular–arbuscular mycorrhizal associations in roots of *Medieago truncatula*. Mol. Plant Microbe Interact. 6, 643–654.

Hart, M.M., Klironomos, J.N., 2002. Diversity of arbuscular mycorrhizal fungi and ecosystem functioning. In: van der Heijden, M.G.A., Sanders, I. (Eds.), Mycorrhizal Ecology. Ecological Studies, vol. 157. Springer-Verlag, Berlin, pp. 225–242.

Hart, M.M., Reader, R.J., 2002a. Taxonomic basis for variation in the colonization strategy of arbuscular mycorrhizal fungi. New Phytol. 153, 335–344.

Hart, M.M., Reader, R.J., 2002b. Host plant benefit from association with arbuscular mycorrhizal fungi: variation due to differences in size of mycelium. Biol. Fertil. Soils 36, 357–366.

Hart, M.M., Reader, R.J., 2005. The role of the external mycelium in early colonization for three arbuscular mycorrhizal fungal species with different colonization strategies. Pedobiologia 49, 269–279.

Hart, M.M., Forsythe, J., Oshowski, B., Bücking, H., Jansa, J., Kiers, E.T., 2013. Hiding in a crowd-does diversity facilitates persistence of a low-quality fungal partner in the mycorrhizal symbiosis. Symbiosis 59, 47–56.

Haug, A., 1984. Molecular aspects of aluminum toxicity. CRC Crit. Rev. Plant Sci. 1, 345–373.

Hause, B., Fester, T., 2005. Molecular and cell biology of arbuscular mycorrhizal symbiosis. Planta 221, 184–196.

Hawkins, H.J., Johansen, A., George, E., 2000. Uptake and transport of organic and inorganic nitrogen by arbuscular mycorrhizal fungi. Plant Soil 226, 275–285.

Hayman, D.S., 1982. Influence of soils and fertility on activity and survival of vesicular-arbuscular mycorrhizal fungi. Phytopathology 72, 1119–1125.

Haystead, A., Malajczuk, N., Grove, T.S., 1988. Underground transfer of nitrogen between pasture plants infected with vesicular-arbuscular mycorrhizal fungi. New Phytol. 108, 417–423.

He, X., Critchley, C., Bledsoe, C., 2003. Nitrogen transfer within and between plants through common mycorrhizal networks (CMNs). CRC Crit. Rev. Plant Sci. 22, 531–567.

He, X., Xu, M., Qiu, G.Y., Zhou, J., 2009. Use of 15N stable isotope to quantify nitrogen transfer between mycorrhizal plants. J. Plant Ecol. 2, 107–118.

Heap, A.J., Newman, E.I., 1980. Links between roots by hyphae of vesicular–arbuscular mycorrhizas. New Physiol. 85, 169–171.

Heckman, J.R., Angle, J.S., 1987. Variation between soybean cultivars in vesicular-arbuscular mycorrhiza fungi colonization. Agron. J. 79, 428–430.

Heggo, A., Angle, J.S., Chaney, R.L., 1990. Effects of vesicular-arbuscular mycorrhizal fungi on heavy metal uptake by soybeans. Soil Biol. Biochem. 22, 865–869.

Helgason, T., Daniell, T.J., Husband, R., Fitter, A., Young, J.P.W., 1998. Ploughing up the wood-wide web? Nature 394, 431.

Helgason, T., Merryweather, J.W., Denison, J., Wilson, P., Young, J.P.W., Fitter, A.H., 2002. Selectivity and functional diversity in arbuscular mycorrhizas of co-occurring fungi and plants from a temperate deciduous woodland. J. Ecol. 90, 371–384.

Helgason, T., Merryweather, J.W., Young, J.P.W., Fitter, A.H., 2007. Specificity and resilience in the arbuscular mycorrhizal fungi of a natural woodland community. J. Ecol. 95, 623–630.

Herrera-Medina, M.J., Steinkellner, S., Vierheilig, H., Ocampo Bote, J.A., Garcia Garrido, J.M., 2007. Abscissic acid determines arbuscule development and functionality in the tomato arbuscular mycorrhiza. New Phytol. 175, 554–564.

Herridge, D.F., Peoples, M.B., Boddey, R.M., 2008. Global inputs of biological nitrogen fixation in agricultural systems. Plant Soil 311, 1–18.

Hetrick, B.A.D., Kitt, D.G., Wilson, G.T., 1988. Mycorrhizal dependence and growth habit of warm-season and cool-season tallgrass prairie plants. Can. J. Bot. 66, 1376–1380.

Hetrick, B.A.D., Wilson, G.W.T., Cox, T.S., 1993. Mycorrhizal dependence of modern wheat cultivars and ancestors: a synthesis. Can. J. Bot. 71, 512–518.

Hibbett, D., 2016. The invisible dimension of fungal diversity. Science 351, 1150–1151. Available from: http://dx.doi.org/10.1126/science.aae0380.

Hiiesalu, I., Opik, M., Metsis, M., Lilje, L., Davison, J., Vasar, M., et al., 2012. Plant species richness belowground: higher richness and new patterns revealed by next-generation sequencing. Mol. Ecol. 21, 2004–2016.

Hiiesalu, I., Partel, M., Davison, J., Gerhold, P., Metsis, M., Moora, M., et al., 2014. Species richness of arbuscular mycorrhizal fungi: associations with grassland plant richness and biomass. New Phytol. 203, 233–244.

Hijri, I., Sy·korová, Z., Oehl, F., Ineichen, K., Mäder, P., Wiemken, A., Redecker, D., 2006. Communities of arbuscular mycorrhizal fungi in arable soils are not necessarily low in diversity. Mol. Ecol. 15, 2277–2289.

Hildermann, I., Messmer, M., Dubois, D., Boller, T., Wiemken, A., Mäder, P., 2010. Nutrient use efficiency and arbuscular mycorrhizal root colonization of winter wheat cultivars in different farming systems of the DOK long-term trial. J. Sci. Food Agric. 90, 2027–2038.

Hirsch, A.M., 1992. Developmental biology of legume nodulation. New Phytol. 122, 211–237.

Hirsch, A.M., Fang, Y., Asad, S., Kapulnik, Y., 1997. The role of phytohormones in plant-microbe symbioses. Plant Soil 194, 171–184.

Hobbs, P.R., Sayre, K., Gupta, R., 2008. The role of conservation agriculture in sustainable agriculture. Philos. Trans. R. Soc. B 363, 543–555.

Hodge, A., 2000. Microbial ecology of the arbuscular mycorrhiza. FEMS Microbiol. Ecol. 32, 91–96.

Hodge, A., Campbell, C.D., Fitter, A.H., 2001. An arbuscular mycorrhizal fungus accelerates decomposition and acquires nitrogen directly from organic material. Nature 413, 297–299.

Hoeksema, J.D., Chaudhary, V.B., Gehring, C.A., Johnson, N.C., Karst, J., Koide, R.T., et al., 2010. A meta-analysis of context dependency in plant response to inoculation with mycorrhizal fungi. Ecol. Lett. 13, 394–407.

Holford, I.C.R., 1997. Soil phosphorus: Its measurement, and its uptake by plants. Aust. J. Soil Res. 35, 227–239.

Holland, T.C., Bowen, P., Bogdanoff, C., Hart, M.M., 2014. How distinct are arbuscular mycorrhizal fungal communities associating with grapevines? Biol. Fertil. Soils 50, 667–674.

Hooper, D.U., Bignell, D.E., Brown, V.K., Brussaard, L., Dangerfield, J.M., Wall, D.H., et al., 2000. Interactions between aboveground and belowground biodiversity in terrestrial ecosystems: patterns, mechanisms, and feedbacks. Bioscience 50, 1049–1061.

Hungate, B.A., Jaeger, C.H., Gamara, G., Chapin, F.S., Field, C.B., 2000. Soil microbiota in two annual grasslands: responses to elevated atmospheric $CO_2$. Oecologia 124, 589–598.

Hunt, H.W., Wall, D.H., 2002. Modelling the effects of loss of soil biodiversity on ecosystem function. Global Change Biol. 8, 33–50.

Hunt, M.E., Floyd, G.L., Stout, B.B., 1979. Soil algae in field and forest environments. Ecology 60, 362–375.

Islas, A.J.T., Eyherabide, M., Echeverría, H.E., Rozas, H.R.S., Covacevich, F., 2014. Mycotrophic capacity and effi ciency of microbial consortia of arbuscular mycorrhizal fungi native of soils from Buenos Aires province under contrasting management. Rev. Argent. Microbiol. 46, 133–143.

Jäderlund, L., Arthurson, V., Granhall, U., Jansson, J.K., 2008. Specific interactions between arbuscular mycorrhizal fungi and plant growth-promoting bacteria: as revealed by different combinations. FEMS Microbiol. Lett. 287, 174–180.

Jakobsen, I., 2004. Hyphal fusion to plant species connections—giant mycelia and community nutrient flow. New Phytol. 164, 4–7.

Jakobsen, I., Hammer, E.C., 2014. Nutrient dynamics in arbuscular mycorrhizal networks. In: Horton, T. (Ed.), Mycorrhizal Networks. Springer, Netherlands, pp. 91–131.

Jakobsen, I., Rosendahl, L., 1990. Carbon flow into soil and external hyphae from roots of mycorrhizal cucumber plants. New Phytol. 115, 77–83.

Jakobsen, I., Abbott, L.K., Robson, A.D., 1992a. External hyphae of vesicular-arbuscular mycorrhizal fungi associated with *Trifolium subterraneum. L.* 1: spread of hyphae and phosphorus inflow into roots. New Phytol. 120, 371–380.

Jakobsen, I., Abbott, L.K., Robson, A.D., 1992b. External hyphae of vesicular-arbuscular mycorrhizal fungi associated with *Trifolium subterraneum. L.*. 2: hyphal transport of $^{32}P$ over defined distances. New Phytol. 120, 509–516.

James, T.Y., Kauff, F., Schoch, C.L., Matheny, P.B., Hofstetter, V., Cox, C.J., et al., 2006. Reconstructing the early evolution of fungi using a six-gene phylogeny. Nature 443, 818–822. Available from: http://dx.doi.org/10.1038/nature05110.

Jansa, J., Erb, A., Oberholzer, H.-R., Šmilauer, P., Egli, S., 2014. Soil and geography are more important determinants of indigenous arbuscular mycorrhizal communities than management practices in Swiss agricultural soils. Mol. Ecol. 23, 2118–2135. Available from: http://dx.doi.org/10.1111/mec.12706.

Jansa, J., Mozafar, A., Anken, T., Ruh, R., Sanders, I.R., Frossard, E., 2002. Diversity and structure of AMF communities as affected by tillage in a temperate soil. Mycorrhiza 12, 225–234.

Jansa, J., Mozafar, A., Kuhn, G., Anken, T., Ruh, R., Sanders, I.R., et al., 2003. Soil tillage affects the community structure of mycorrhizal fungi in maize roots. Ecol. Appl. 13, 1164–1176.

Jansa, J., Mozafar, A., Frossard, E., 2005. Phosphorus acquisition strategies within arbuscular mycorrhizal fungal community of a single field site. Plant Soil 276, 163–176.

Jansa, J., Smith, F.A., Smith, S.E., 2008. Are there benefits of simultaneous root colonization by different arbuscular mycorrhizal fungi? New Phytol. 177, 779–789. Available from: http://dx.doi.org/10.1111/j.1469-8137.2007.02294.x.

Janssen, B.H., de Willigen, P., 2006. Ideal and saturated soil fertility as bench marks in nutrient management 1. Outline of the framework. Ecosyst. Environ. 116, 132–146.

Jasper, D.A., Abbott, L.K., Robson, A.D., 1989a. Soil disturbance reduces the infectivity of external hyphae of vesicular–arbuscular mycorrhizal fungi. New Phytol. 112, 93–99.

Jasper, D.A., Abbott, L.K., Robson, A.D., 1989b. Hyphae of a vesicular–arbuscular mycorrhizal fungus maintain infectivity in dry soil, except when the soil is disturbed. New Phytol. 112, 101–107.

Jasper, D.A., Abbott, L.K., Robson, A.D., 1991. The effect of soil disturbance on vesicular–arbuscular mycorrhizal fungi in soils from different vegetation types. New Phytol. 118, 471–476.

Jasper, D.A., Abbott, L.K., Robson, A.D., 1993. The survival of infective hyphae of vesicular–arbuscular mycorrhizal fungi in dry soil: an interaction with sporulation. New Phytol. 124, 473–479.

Jasper, D.A., Robson, A.D., Abbott, L., 1979. Phosphorus and the formation of vesicular-arbuscular mycorrhizas. Soil Biol. Biochem. 11, 501–505.

Jayachandran, K., Schwab, A.P., Hetrick, B.A.D., 1989. Mycorrhizal mediation of phosphorus availability: synthetic iron chelate effects on phosphorus solubilization. Soil Sci. Soc. Am. J. 53, 1701–1706.

Jayachandran, K., Schwab, A.P., Hetrick, B.A.D., 1992. Mineralization of organic phosphorus by vesicular–arbuscular mycorrhizal fungi. Soil Biol. Biochem. 24, 897–903.

Jeffries, P., Barea, J.M., 2012. Arbuscular mycorrhiza: a key component of sustainable plant-soil ecosystems. In: Hock, B. (Ed.), The *Mycota*, a Comprehensive Treatise on Fungi as Experimental Systems for Basic and Applied Research. Springer, Berlin, Heidelberg, pp. 51–75.

Jin, H., Pfeffer, P.E., Douds, D.D., Piotrowski, E., Lammers, P.J., Shachar-Hill, Y., 2005. The uptake, metabolism, transport and transfer of nitrogen in an arbuscular mycorrhizal symbiosis. New Phytol. 168, 687–696.

Johansen, A., Jensen, E.S., 1996. Transfer of N and P from intact or decomposing roots of pea to barley interconnected by an arbuscular mycorrhizal fungus. Soil Biol. Biochem. 28, 73–81.

Johnson, N.C., 1993. Can fertilization of the soil select less mutualistic mycorrhizae? Ecol. Appl. 3, 749–757.

Johnson, D., Gilbert, L., 2015. Interplant signalling through hyphal networks. New Phytol. 205, 1448–1453.

Johnson, D., Vandenkoornhuyse, P.J., Leake, J.R., Gilbert, L., Booth, R.E., Grime, J.P., et al., 2003. Plant communities affect arbuscular mycorrhizal fungal diversity and community composition in grassland microcosms. New Phytol. 161, 503–515.

Johnson, C.K., Wienhold, B.J., Doran, J.W., Drijber, R.A., Wright, S.F., 2004. Linking microbial-scale findings to farm-scale outcomes in a dryland cropping system. Precis. Agric. 5, 311–328.

Johnson, N.C., Pfleger, F.L., 1992. Vesicular-arbuscular mycorrhizae and cultural stress. In: Bethlenfalvay, G.J., Linderman, R.G. (Eds.), Mycorrhizae in Sustainable Agriculture. American Society of Agronomy, Madison, WI, pp. 71–99. ASA/CSSA/SSSA Special Publication No. 54.

Johnson, N.C., Copeland, P.J., Crookston, R.K., Pfleger, F.L., 1992. Mycorrhizae: possible explanation for yield decline with continuous corn and soybean. Agron. J. 84, 387–390.

Johnson, N.C., Graham, J.H., Smith, F.A., 1997. Functioning of mycorrhizal associations along the mutualism-parasitism continuum. New Phytol. 135, 575–586.

Jones, K.M., Walker, G.C., 2008. Responses of the model legume *Medicago truncatula* to the rhizobial exopolysaccharide succinoglycan. Plant Signaling Behav. 3, 888–890.

Juniper, S., Abbott, L.K., 2006. Soil salinity delays germination and limits growth of hyphae from propagules of arbuscular mycorrhizal fungi. Mycorrhiza 16, 371–379.

Jun-Li, H., Xian-Gui, L., Jun-Hua, W., Wei-Shou, S., Shu, W., Su-Ping, P., et al., 2010. Arbuscular mycorrhizal fungal inoculation enhances suppression of cucumber Fusarium wilt in greenhouse soils. Pedosphere 20, 586–593.

Kabir, Z., 2005. Tillage or no-tillage: impact on mycorrhizae. Can. J. Plant Sci. 85, 23–29.

Kabir, Z., Koide, R.T., 2000. The effect of dandelion or a cover crop on mycorrhiza inoculum potential, soil aggregation and yield of maize. Agric. Ecosyst. Environ. 78, 167–174.

Kabir, Z., Koide, R.T., 2002. Effect of autumn and winter mycorrhizal cover crops on soil properties, nutrient uptake and yield of sweet corn in Pennsylvania, USA. Plant Soil 238, 205–215.

Kabir, Z., O'Halloran, I.P., Fyles, J.W., Hamel, C., 1997a. Seasonal changes of arbuscular mycorrhizal fungi as affected by tillage practices and fertilization: hyphal density and mycorrhizal root colonization. Plant Soil 192, 285–293.

Kabir, Z., O'Halloran, I.P., Hamel, C., 1997b. Overwinter survival of arbuscular mycorrhizal hyphae is favoured by attachment to roots but diminished by disturbance. Mycorrhiza 7, 197–200.

Kabir, Z., O'Halloran, I.P., Widden, P., Hamel, C., 1998. Vertical distribution of arbuscular mycorrhizal fungi under corn (*Zea mays* L.) in no-till and conventional tillage systems. Mycorrhiza 8, 53–55.

Kadir, S., 1994. Interaction Between Infection With Rhizobium and Indigenous Mycorrhizas on Nitrogen Fixation and Phosphorus Nutrition in Soybean. M.Sc. Thesis. University of Guelph, Guelph, Ontario, Canada.

Kahiluoto, H., 2004. Systems management of AM in sustainable agriculture the case of P supply. In: Baar, J., Josten, E. (Eds.), Role of Mycorrhiza in Sustainable Land Management. Programme and Abstracts of the COST-Meeting: Role of Mycorrhiza in Sustainable Land Management.

Kahiluoto, H., Ketoja, E., Vestberg, M., 2000. Promotion of utilization of arbuscular mycorrhiza through reduced P fertilization 1. Bioassays in a growth chamber. Plant Soil 227, 191–206.

Kahiluoto, H., Ketoja, E., Vestberg, M., Saarela, I., 2001. Promotion of AM utilization through reduced P fertilization 2. Field studies. Plant Soil 231, 65–79.

Kahiluoto, H., Ketoja, E., Vestberg, M., 2009. Contribution of arbuscular mycorrhiza to soil quality in contrasting cropping systems. Agric. Ecosyst. Environ. 134, 36–45.

Kahiluoto, H., Ketoja, E., Vestberg, M., 2012. Plant-available P supply is not the main factor determining the benefit from arbuscular mycorrhiza to crop P nutrition and growth in contrasting cropping systems. Plant Soil 350, 85–98.

Kapulnik, Y., Tsror, L., Zipori, I., Hazanovsky, M., Wininger, S., Dag, A., 2010. Effect of AMF application on growth, productivity and susceptibility to verticillium wilt of olives grown under desert conditions. Symbiosis 52, 103–111.

Karasawa, T., Arihara, J., Kasahara, Y., 2000a. Effects of previous crops on arbuscular mycorrhizal formation and growth of maize under various soil moisture conditions. Soil Sci. Plant Nutr. 46, 53–60.

Karasawa, T., Takebe, M., Kasahara, Y., 2000b. Arbuscular mycorrhizal (AM) effects on maize growth and am colonization of roots under various soil moisture conditions. Soil Sci. Plant Nutr. 46, 61–67.

Karasawa, T., Kasahara, Y., Takebe, A., 2002. Differences in growth responses of maize to preceding cropping caused by fluctuation in the population of indigenous arbuscular mycorrhizal fungi. Soil Biol. Biochem. 34, 851–857.

Kaschuk, G., Kuyper, T.W., Leffelaar, P.A., Hungria, M., Giller, K.E., 2009. Are the rates of photosynthesis stimulated by the carbon sink strength of rhizobial and arbuscular mycorrhizal symbioses? Soil Biol. Biochem. 41, 1233–1244.

Kassam, A., Brammer, H., 2013. Combining sustainable agricultural production with economic and environmental benefits. Geogr. J. 179, 11–18.

Kayombo, B., Lal, R., 1994. Response of tropical crops to soil compaction. In: Sloane, B.D., Van Ouwerkkerk, C. (Eds.), Soil Compaction in Crop Production. Elsevier, Amsterdam, pp. 287–315.

Kefi, M., Yoshino, K., 2010. Evaluation of the economic effects of soil erosion risk on agricultural productivity using remote sensing: case of watershed in Tunisia. International Archives of the Photogrammetry, Remote Sensing and Spatial Information Science 37, 930–935.

Khalil, S., Loynachan, T.E., McNabb, H.S., 1992. Colonization of soybean by mycorrhizal fungi and spore populations in Iowa soils. Agron. J. 84, 832–836.

Khalvati, M.A., Hu, Y., Mozafar, A., Schmidhalter, U., 2005. Quantification of water uptake by arbuscular mycorrhizal hyphae and its significance for leaf growth, water relations, and gas exchange of barley subjected to drought Stress. Plant Biol. 7, 706–712.

Khaosaad, T., García Garrido, J.M., Steinkellner, S., Vierheilig, H., 2007. Take-all disease is systemically reduced in roots of mycorrhizal barley plants. Soil Biol. Biochem. 39, 727–734.

Kiers, E.T., Duhamel, M., Beesetty, Y., Mensah, J.A., Franken, O., Verbruggen, E., et al., 2011. Reciprocal rewards stabilize cooperation in the mycorrhizal symbiosis. Science 333, 880–882.

Kikuchi, Y., Hijikata, N., Ohtomo, R., Handa, Y., Kawaguchi, M., Saito, K., et al., 2016. Aquaporin-mediated long-distance polyphosphate translocation directed towards the host in arbuscular mycorrhizal symbiosis: application of virus-induced gene silencing. New Phytol. 211, 1202–1208.

Kirk, A.P., Entz, M.H., Fox, S.L., Tenuta, M., 2011. Mycorrhizal colonization, P uptake and yield of older and modern wheats under organic management. Can. J. Plant Sci. 91, 663–667.

Kistner, C., Parniske, M., 2002. Evolution of signal transduction in intracellular symbiosis. Trends Plant Sci. 7, 511–518.

Klironomos, J.N., 2003. Variation in plant response to native and exotic arbuscular mycorrhizal fungi. Ecology 84, 2292–2301.

Klironomos, J.N., Hart, M.M., 2002. Colonization of roots by arbuscular mycorrhizal fungi using different sources of inoculum. Mycorrhiza 12, 181–184.

Klironomos, J., Ursic, M., 1998. Density-dependent grazing on the extraradical hyphal network of the arbuscular mycorrhizal fungus, *Glomus intraradices*, by the collembolan, *Folsomia candida*. Biol. Fertil. Soils 26, 250–253.

Klironomos, J., Zobel, M., Tibbett, M., Stock, W.D., Rillig, M.C., Parrent, J.L., et al., 2011. Forces that structure plant communities: quantifying the importance of the mycorrhizal symbiosis. New Phytol. 189, 366–370.

Kobra, N., Jalil, K., Youbert, G., 2009. Effects of three *Glomus* species as biocontrol agents against verticillium-induced wilt in cotton. J. Plant Protect. Res. 49, 185–189.

Koch, A.M., Croll, D., Sanders, I.R., 2006. Genetic variability in a population of arbuscular mycorrhizal fungi causes variation in plant growth. Ecol. Lett. 9, 103–110.

Koch, A.M., Kuhn, G., Fontanillas, P., Fumagalli, L., Goudet, J., Sanders, I.R., 2004. High genetic variability and low local diversity in a population of arbuscular mycorrhizal fungi. Proc. Natl. Acad. Sci. USA 101, 2369–2374.

Köhl, L., Oehl, F., van der Heijden, M.G.A., 2014. Agricultural practices indirectly influence plant productivity and ecosystem services through effects on soil biota. Ecol. Appl. 24, 1842–1853.

Kohout, P., Sudová, R., Janoušková, M., Čtvrtlíková, M., Hejda, M., Pánková, H., et al., 2014. Comparison of commonly used primer sets for evaluating arbuscular mycorrhizal fungal communities: Is there a universal solution? Soil Biol. Biochem. 68, 482–493. Available from: http://dx.doi.org/10.1016/j.soilbio.2013.08.027.

Koide, R.T., 2000. Functional complementarity in the arbuscular mycorrhizal symbiosis. New Phytol. 147, 233–235.

Koide, R.T., Kabir, Z., 2000. Extraradical hyphae of the mycorrhizal fungus *Glomus intraradices* can hydrolyze organic phosphate. New Phytol. 148, 511–517.

Kõljalg, U., Nilsson, R.H., Abarenkov, K., Tedersoo, L., Taylor, A.F.S., Bahram, M., et al., 2013. Towards a unified paradigm for sequence-based identification of fungi. Mol. Ecol. 22, 5271–5277. Available from: http://dx.doi.org/10.1111/mec.12481.

Kothari, S.K., Marschner, H., Römheld, V., 1990. Direct and indirect effects of VA mycorrhizial fungi and rhizosphere microorganisms on acquisition of mineral nutrients by maize (Zea mays) in a calcareous soil. New Phytol. 116, 637–645.

Kothari, S.K., Marschner, H., Romheld, V., 1991. Effect of a vesicular arbuscular mycorrhizal fungus and rhizosphere microorganisms on manganese reduction in the rhizosphere and manganese concentrations in maize (*Zea mays* L). New Phytol. 117, 649–655.

Kowalchuk, G.A., De Souza, F.A., Van Veen, J.A., 2002. Community analysis of arbuscular mycorrhizal fungi associated with *Ammophila arenaria* in Dutch coastal sand dunes. Mol. Ecol. 11, 571–581.

Kowalchuk, G.A., Drigo, B., Yergeau, E., van Veen, J.A., 2006. Assessing bacterial and fungal community structure in soil using ribosomal rna and other structural gene markers. Nucleic Acids and Proteins in Soil, Soil Biology. Springer Berlin Heidelberg, Berlin, Heidelberg, pp. 159–188. Available from: http://dx.doi.org/10.1007/3-540-29449-X_8.

Kreuzer, K., Adamczyk, J., Iijima, M., Wagner, M., Scheu, S., Bonkowski, M., 2006. Grazing of a common species of soil protozoa (*Acanthamoeba castellanii*) affects rhizosphere bacterial community composition and root architecture of rice (*Oryza sativa* L.). Soil Biol. Biochem. 38, 1665–1672.

Krikun, J., 1991. Mycorrhizae in agricultural crops. In: Waisel, Y., Eshel, A., Kafkafi, U. (Eds.), Plant Roots, The Hidden Half. Marcel Dekker, New York, pp. 767–788.

Krishnamoorthy, R., Kim, K., Kim, C., Sa, T., 2014. Changes of arbuscular mycorrhizal traits and community structure with respect to soil salinity in a coastal reclamation land. Soil Biol. Biochem. 72, 1–10.

Krol, E., Becker, A., 2004. Global transcriptional analysis of the phosphate starvation response in *Sinorhizobium meliloti* strains 1021 and 2011. Mol. Genet. Genomics 272, 1–17.

Kruger, M., Kruger, C., Walker, C., Stockinger, H., Schussler, A., 2012. Phylogenetic reference data for systematics and phylotaxonomy of arbuscular mycorrhizal fungi from phylum to species level. New Phytol. 193, 970–984.

Krüger, M., Stockinger, H., Krüger, C., Schüßler, A., 2009. DNA-based species level detection of Glomeromycota: one PCR primer set for all arbuscular mycorrhizal fungi. New Phytol. 183, 212–223. Available from: http://dx.doi.org/10.1111/j.1469-8137.2009.02835.x.

Krüger, M., Teste, F.P., Laliberté, E., Lambers, H., Coghlan, M., Zemunik, G., et al., 2015. The rise and fall of arbuscular mycorrhizal fungal diversity during ecosystem retrogression. Mol. Ecol. 24, 4912–4930. Available from: http://dx.doi.org/10.1111/mec.13363.

Kucey, R.M.N., Paul, E.A., 1981. Carbon flow in plant microbial associations. Science, New Series 213, 473–474.

Kucey, R.M.N., Paul, E.A., 1982. Carbon flow, photosynthesis, and $N_2$ fixation in mycorrhizal and nodulated faba beans (*Vicia faba* L.). Soil Biol. Biochem. 14, 407–412.

Lal, R., 1998. Soil erosion impact on agronomic productivity and environment quality. CRC Crit. Rev. Plant Sci. 17, 319–464.

Lal, R., 2015. Sequestering carbon and increasing productivity by conservation agriculture. J. Soil Water Conserv. 70, 55–62.

Landis, F.C., Gargas, A., Givnish, T.J., 2004. Relationships among arbuscular mycorrhizal fungi, vascular plants and environmental conditions in oak savannas. New Phytol. 164, 493–504.

Lane, D.J., Pace, B., Olsen, G.J., Stahl, D.A., Sogin, M.L., Pace, N.R., 1985. Rapid determination of 16S ribosomal RNA sequences for phylogenetic analyses. Proc. Natl. Acad. Sci. USA 82, 6955–6959.

Lanfranco, L., Delpero, M., Bonfante, P., 1999. Intrasporal variability of ribosomal sequences in the endomycorrhizal fungus *Gigaspora margarita*. Mol. Ecol. 8, 37–45.

Larimer, A.L., Bever, J.D., Clay, K., 2010. The interactive effects of plant microbial symbionts: a review and metaanalysis. Symbiosis 51, 139–148.

Larimer, A.L., Clay, K., Bever, J.D., 2014. Synergism and context dependency of interactions between arbuscular mycorrhizal fungi and rhizobia with a prairie legume. Ecology 95, 1045–1054.

Larsen, J., Jakobsen, I., 1996. Interactions between a mycophagous Collembola, dry yeast and the external mycelium of an arbuscular mycorrhizal fungus. Mycorrhiza 6, 259–264.

Larsen, J., Cornejo, P., Barea, J.M., 2009. Interactions between the arbuscular mycorrhizal fungus *Glomus intraradices* and the plant growth promoting rhizobacteria *Paenibacillus polymyxa* and *P. macerans* in the mycorrhizosphere of *Cucumis sativus*. Soil Biol. Biochem. 41, 286–292.

Law, I.J., Strijdom, B.W., 1984. Role of lectins in the specific recognition of rhizobium by *Lotononis bainesii*. Plant Physiol. 74, 779–785.

Lax, P., Becerra, A.G., Soteras, F., Cabello, M., Doucet, M.E., 2011. Effect of the arbuscular mycorrhizal fungus Glomus intraradices on the false root-knot nematode *Nacobbus aberrans* in tomato plants. Biol. Fertil. Soils 47, 591–597.

Leake, J., Johnson, D., Donnelly, D., Muckle, G., Boddy, L., Read, D., 2004. Networks of power and influence: the role of mycorrhizal mycelium in controlling plant communities and agroecosystem functioning. Can. J. Bot. 82, 1016–1045.

Lee, J.-E., Eom, A.-H., 2009. Effect of organic farming on spore diversity of arbuscular mycorrhizal fungi and glomalin in soil. Mycobiology 37, 272–276.

Lee, P.-J., Koske, R.E., 1994. *Gigaspora gigantea*: parasitism of spores by fungi and actinomycetes. Mycol. Res. 98, 458–466.

Lee, J., Lee, S., Young, J.P.W., 2008. Improved PCR primers for the detection and identification of arbuscular mycorrhizal fungi. FEMS Microbiol. Ecol. 65, 339–349. Available from: http://dx.doi.org/10.1111/j.1574-6941.2008.00531.x.

Lee, E.H., Eo, J.K., Ka, K.H., Eom, A.H., 2013. Diversity of arbuscular mycorrhizal fungi and their roles in ecosystems. Mycobiology 41, 121–125.

Lehman, R.M., Taheri, W.I., Osborne, S.L., Buyer, J.S., Douds Jr, D.D., 2012. Fall cover cropping can increase arbuscular mycorrhizae in soils supporting intensive agricultural production. Appl. Soil Ecol. 61, 300–304.

Lehmann, A., Veresoglou, S.D., Leifheit, E.F., Rillig, M.C., 2014. Arbuscular mycorrhizal influence on Zinc nutrition in crop plants—a meta-analysis. Soil Biol. Biochem. 69, 123–131.

Lekberg, Y., Koide, R.T., 2005. Is plant performance limited by an abundance of arbuscular mycorrhizal fungi? A meta-analysis of studies published between 1988-2003. New Phytol. 168, 189–204.

Lekberg, Y., Gibbons, S.M., Rosendahl, S., Ramsey, P.W., 2013. Severe plant invasions can increase mycorrhizal fungal abundance and diversity. ISME J. 7, 1424–1433.

Lekberg, Y., Gibbons, S.M., Rosendahl, S., 2014. Will different OTU delineation methods change interpretation of arbuscular mycorrhizal fungal community patterns? New Phytol. 202, 1101–1104.

Lendzemo, V.W., Kuyper, T.W., 2001. Effects of arbuscular mycorrhizal fungi on damage by *Striga hermonthica* on two contrasting cultivars of sorghum, Sorghum bicolor. Agric. Ecosyst. Environ. 87, 29–35.

Leteinturier, B., Herman, J.L., de Longueville, F., Quintin, L., Oger, R., 2006. Adaptation of a crop sequence indicator based on a land parcel management system. Agric. Ecosyst. Environ. 112, 324–334.

Leveau, J.H.J., Preston, G.M., 2008. Bacterial mycophagy: definition and diagnosis of a unique bacterial–fungal interaction. New Phytol. 177, 859–876.

Lewis, J.A., Papavizas, G.C., 1991. Biocontrol of plant diseases—the approach for tomorrow. Crop Prot. 10, 95–105.

Li, H., Wang, C., Li, X., Xiang, D., 2013. Inoculating maize fields with earthworms (*Aporrectodea trapezoides*) and an arbuscular mycorrhizal fungus (*Rhizophagus intraradices*) improves mycorrhizal community structure and increases plant nutrient uptake. Biol. Fertil. Soils 49, 1167–1178.

Li, X.L., George, E., Marchner, H., 1991. Extension of the phosphorus depletion zone in VA-mycorrhizal white clover in a calcareous soil. Plant Soil 136, 41–48.

Lin, K., Limpens, E., Zhang, Z., Ivanov, S., Saunders, D.G.O., Mu, D., et al., 2014. Single nucleus genome sequencing reveals high similarity among nuclei of an endomycorrhizal fungus. PLoS Genet. 10, e1004078. Available from: http://dx.doi.org/10.1371/journal.pgen.1004078.s022.

Lioussanne, L., Perreault, F., Jolicoeur, M., St-Arnaud, M., 2010. The bacterial community of tomato rhizosphere is modified by inoculation with arbuscular mycorrhizal fungi but unaffected by soil enrichment with mycorrhizal root exudates or inoculation with *Phytophthora nicotianae*. Soil Biol. Biochem. 42, 473–483.

Liu, A., Hamel, C., Hamilton, R.I., Ma, B.L., Smith, D.L., 2000. Acquisition of Cu, Zn, Mn and Fe by mycorrhizal maize (*Zea mays* L.) grown in soil at different P and micronutrient levels. Mycorrhiza 9, 331–336.

Liu, K.-L., Porras-Alfaro, A., Kuske, C.R., Eichorst, S.A., Xie, G., 2012a. Accurate, rapid taxonomic classification of fungal large-subunit rRNA genes. Appl. Environ. Microbiol. 78, 1523–1533. Available from: http://dx.doi.org/10.1128/AEM.06826-11.

Liu, Y.J., Shi, G.X., Mao, L., Cheng, G., Jiang, S.J., Ma, X.J., et al., 2012b. Direct and indirect influences of 8 yr of nitrogen and phosphorus fertilization on Glomeromycota in an alpine meadow ecosystem. New Phytol. 194, 523–535.

Liu, S., Guo, X., Feng, G., Maimaitiaili, B., Fan, J., He, X., 2016. Indigenous arbuscular mycorrhizal fungi can alleviate salt stress and promote growth of cotton and maize in saline fields. Plant Soil 398, 195–206.

Liu, T., Sheng, M., Wang, C.Y., Chen, H., Li, Z., Tang, M., 2015. Impact of arbuscular mycorrhizal fungi on the growth, water status, and photosynthesis of hybrid poplar under drought stress and recovery. Photosynthetica 53, 250–258.

Logi, C., Sbrana, C., Giovannetti, M., 1998. Cellular events involved in survival of individual arbuscular mycorrhizal symbionts growing in the absence of the host. Appl. Environ. Microbiol. 64, 3473–3479.

Long, S.R., 1996. Rhizobium symbiosis: nod factors in perspective. Plant Cell 8, 1885–1898.

Lu, S., Braunberger, P.G., Miller, M.H., 1994. Response of vesicular-arbuscular mycorrhizas of maize to various rates of P addition to different rooting zones. Plant Soil 158, 119–128.

Lumini, E., Bianciotto, V., Jargeat, P., Novero, M., Salvioli, A., Faccio, A., et al., 2007. Presymbiotic growth and sporal morphology are affected in the arbuscular mycorrhizal fungus *Gigaspora margarita* cured of its endobacteria. Cell Microbiol. 9, 1716–1729.

Lumini, E., Orgiazzi, A., Borriello, R., Bonfante, P., Bianciotto, V., 2010. Disclosing arbuscular mycorrhizal fungal biodiversity in soil through a land-use gradient using a pyrosequencing approach. Environ. Microbiol. 12, 2165–2179.

MacDonald, R.M., Chandler, M.R., 1981. Bacterium-like organelles in the vesicular-arbuscular, mycorrhizal fungus *Glomus catedonius*. New Phytol. 89, 241–246.

MacDonald, R.M., Chandler, M., Mosse, B., 1982. The occurrence of bacterium-like organelles in vesicular-arbuscular mycorrhizal fungi. New Phytol. 90, 659–663.

Mäder, P., Vierheilig, H., Streitwolf-Engel, R., Boller, T., Frey, B., Christie, P., et al., 2000. Transport of $^{15}N$ from a soil compartment separated by a polytetrafluoroethylene membrane to plant roots via the hyphae of arbuscular mycorrhizal fungi. New Phytol. 146, 155–161.

Madsen, E.L., 2005. Identifying microorganisms responsible for ecologically significant biogeochemical processes. Nat. Rev. Microbiol. 3, 439–446.

Maffei, G., Miozzi, L., Fiorilli, V., Novero, M., Lanfranco, L., Accotto, G.P., 2014. The arbuscular mycorrhizal symbiosis attenuates symptom severity and reduces virus concentration in tomato infected by Tomato yellow leaf curl Sardinia virus (TYLCSV). Mycorrhiza 24, 179–186.

Maherali, H., Klironomos, J., 2007. Influence of phylogeny on fungal community assembly and ecosystem functioning. Science 316, 1746–1748.

Maherali, H., Klironomos, J.N., 2012. Phylogenetic and trait-based assembly of arbuscular mycorrhizal fungal communities. PLoS ONE 7 (5), e36695.

Maillet, F., Poinsot, V., André, O., Puech-Pagés, V., Haouy, A., Gueunier, M., et al., 2011. Fungal lipochitooligosaccharide symbiotic signals in arbuscular mycorrhiza. Nature 469, 58–63.

Maldonado-Mendoza, I.E., Dewbre, G.R., Harrison, M.J., 2001. A phosphate transporter gene from the extra-radical mycelium of an arbuscular mycorrhizal fungus *Glomus intraradices* is regulated in response to phosphate in the environment. Mol. Plant Microbe Interact. 14, 1140–1148.

Mandyam, K., Jumpponen, A., 2005. Seeking the elusive function of the root-colonising dark septate endophytic fungi. Stud. Mycol. 53, 173–189.

Mao, L., Liu, Y.J., Shi, G.X., Jiang, S.J., Cheng, G., Li, X.M., et al., 2014. Wheat cultivars form distinctive communities of root-associated arbuscular mycorrhiza in a conventional agroecosystem. Plant Soil 374, 949–961.

Martinez, T.N., Johnson, N.C., 2010. Agricultural management influences propagule densities and functioning of arbuscular mycorrhizas in low- and high-input agroecosystems in arid environments. Appl. Soil Ecol. 46, 300–306.

Martinez-Garcia, L.B., Garcia, K., Hammer, E.C., Vayssieres, A., 2013. Mycorrhiza for all: an under-earth revolution. New Phytol. 198, 652–655.

Martins, M.A., Read, D.J., 1997. The effects of disturbance on the external mycelium of arbuscular mycorrhizal fungi on plant growth. Pesquisa Agropecuaria Brasileira 32, 1183–1189.

Marulanda, A., Azcón, R., Ruiz-Lozano, J.M., 2003. Contribution of six arbuscular mycorrhizal fungal isolates to water uptake by *Lactuca sativa* plants under drought stress. Physiol. Plant 119, 526–533.

Matekwor Ahulu, E., Gollotte, A., Gianinazzi-Pearson, V., Nonaka, M., 2006. Cooccurring plants forming distinct arbuscular mycorrhizal morphologies harbor similar AM fungal species. Mycorrhiza 17, 37–49. Available from: http://dx.doi.org/10.1007/s00572-006-0079-0.

Maya, M.A., Matsubara, Y., 2013. Influence of arbuscular mycorrhiza on the growth and antioxidative activity in cyclamen under heat stress. Mycorrhiza 23, 381–390.

Mazzoncini, M., Bahadur Sapkota, T., Bàrberi, P., Antichi, D., Risaliti, R., 2011. Long-term effect of tillage, nitrogen fertilization and cover crops on soil organic carbon and total nitrogen content. Soil Tillage Res. 114, 165–174.

McCarter, J.P., 2009. Molecular approaches toward resistance to plant-parasitic nematodes. In: Berg, R.H., Taylor, C.G. (Eds.), Cell Biology of Plant Nematode Parasitism. Plant Cell Monographs. Springer, pp. 239–268.

McGee, P.A., Pattinson, G.S., Heath, R.A., Newman, C.A., Allen, S.J., 1997. Survival of propagules of arbuscular mycorrhizal fungi in soils in eastern Australia used to grow cotton. New Phytol. 135, 773–780.

McGonigle, T.P., 1988. A numerical analysis of published field trials with vesicular-arbuscular mycorrhizal fungi. Funct. Ecol. 2, 473–478.

McGonigle, T.P., Fitter, A.H., 1988. Ecological consequences of arthropod grazing on VA mycorrhizal fungi. Proc. R. Soc. Edinburgh 94B, 25–32.

McGonigle, T.P., Fitter, A.H., 1990. Ecological specificity of vesicular-arbuscular mycorrhizal associations. Mycol. Res. 94, 120–122.

McGonigle, T.P., Miller, M.H., 1993. Mycorrhizal development and phosphorous in maize under conventional and reduced tillage. Soil. Sci. Soc. Am. J. 57, 1002–1006.

McGonigle, T.P., Miller, M.H., 1996. Mycorrhizae phosphorus absorption and yield of maize in response to tillage. Soil Sci. Soc. Am. J. 60, 1856–1861.

McGonigle, T.P., Miller, M.H., 1999. Winter survival of extraradical mycelium and spores of arbuscular mycorrhizal fungi in the field. Appl. Soil Ecol. 12, 41–50.

McGonigle, T.P., Miller, M.H., 2000. The inconsistent effect of soil disturbance on colonization of roots by arbuscular mycorrhizal fungi a test of the inoculum density hypothesis. Appl. Soil Ecol. 14, 147–155.

McGonigle, T.P., Miller, M.H., Evans, D.G., Fairchild, G.L., Swan, J., 1990. A new method which gives an objective measure of colonization of roots by vesicular-arbuscular mycorrhizal fungi. New Phytol. 115, 495–501.

Meddad-Hamza, A., Beddiar, A., Gollotte, A., Lemoine, M.C., Kuszala, C., Gianinazzi, S., 2010. Arbuscular mycorrhizal fungi improve the growth of olive trees and their resistance to transplantation stress. Afr. J. Biotechnol. 9, 1159–1167.

Meier, S., Cornejo, P., Cartes, P., Borie, F., Medina, J., Azcon, R., 2015. Interactive effect between Cu-adapted arbuscular mycorrhizal fungi and biotreated agrowaste residue to improve the nutritional status of *Oenothera picensis* growing in Cu-polluted soils. J. Plant Nutr. Soil Sci. 178, 126–135.

Mena-Violante, H.G., Ocampo-Jimenez, O., Dendooven, L., Martinez-Soto, G., Gonzalez-Castaneda, J., Davies, F.T., et al., 2006. Arbuscular mycorrhizal fungi enhance fruit growth and quality of chile ancho (*Capsicum annuum* L. cv San Luis) plants exposed to drought. Mycorrhiza 16, 261–267.

Mendes, R., Kruijt, M., de Bruijn, I., Dekkers, E., van der Voort, M., Schneider, J.H.M., et al., 2011. Deciphering the rhizosphere microbiome for disease-suppressive bacteria. Science 332, 1097–1100.

Mensah, J.A., Koch, A.M., Antunes, P.M., Kiers, E.T., Hart, M., Bücking, H., 2015. High functional diversity within species of arbuscular mycorrhizal fungi is associated with differences in phosphate and nitrogen uptake and fungal phosphate metabolism. Mycorrhiza 25, 533–546.

Merryweather, J.W., 2001. Comment: meet the glomales—the ecology of mycorrhiza. Br. Wildl. 13, 86–93.

Merryweather, J., Fitter, A., 1998. The arbuscular mycorrhizal fungi of *Hyacinthoides non-scripta*—I. Diversity of fungal taxa. New Phytol. 138, 117–129.

Michot, B., Bachellerie, J.P., 1987. Comparisons of large subunit rRNAs reveal some eukaryote-specific elements of secondary structure. Biochimie 69, 11–23.

Mickan, B., 2014. Mechanisms for alleviation of plant water stress involving arbuscular mycorrhizas. In: Solaiman, Z., Abbott, L., Varma, A. (Eds.), Mycorrhizal Fungi: Use in Sustainable Agriculture and Land Restoration. Springer, Berlin Heidelberg, Germany, pp. 225–239.

Mikkelsen, B.L., Rosendahl, S., Jakobsen, I., 2008. Underground resource allocation between individual networks of mycorrhizal fungi. New Phytol. 180, 890–898.

Millar, N.S., Bennett, A.E., 2016. Stressed out symbiotes: hypotheses for the influence of abiotic stress on arbuscular mycorrhizal fungi. Oecologia 182, 625–641.

Millennium Ecosystem Assessment, 2005. Ecosystems and Human Well-Being: Biodiversity Synthesis. World Resources Institute, Washington, DC.

Miller, M.H., 2000. Arbuscular mycorrhizae and the phosphorus nutrition of maize: a review of Guelph studies. Can. J. Plant Sci. 80, 47–52.

Miller, R.M., Jastrow, J.D., 1990. Hierarchy of root and mycorrhizal fungal interactions with soil aggregation. Soil Biol. Biochem. 22, 579–584.

Miller, R., Jastrow, J., 2000. Mycorrhizal fungi influence soil structure. In: Kapulnik, Y., Douds, D.D. (Eds.), Arbuscular Mycorrhizae: Molecular Biology and Physiology. Kluwer Academic Press, Dordrecht, pp. 3–18.

Miller, R.M., Kling, M., 2000. The importance of integration and scale in the arbuscular mycorrhizal symbiosis. Plant Soil 226, 295–309.

Miller, M.H., McGonigle, T.P., Addy, H.D., 1995a. Functional ecology of vesicular arbuscular mycorrhizas as influenced by phosphate fertilization and tillage in an agricultural ecosystem. Crit. Rev. Biotechnol. 15, 241–255.

Miller, R.M., Reinhardt, D.R., Jastrow, J.D., 1995b. External hyphal production of vesicular arbuscular mycorrhizal fungi in pasture and tallgrass prairie communities. Oecologia 103, 17–23.

Miller, R.M., Miller, S.P., Jastrow, J.D., Rivetta, C.B., 2002. Mycorrhizal mediated feedbacks influence net carbon gain and nutrient uptake in *Andropogon gerardii*. New Phytol. 155, 149–162.

Miranda, J.C.C., Harris, R.J., 1994. The effect of soil phosphorus on the external mycelium growth of arbuscular mycorrhizal fungi during the early stages of mycorrhiza formation. Plant Soil 166, 271–280.

Miransari, M., 2010. Contribution of arbuscular mycorrhizal symbiosis to plant growth under different types of soil stress. Plant Biol. 12, 563–569.

Miransari, M., 2011a. Arbuscular mycorrhizal fungi and nitrogen uptake. Arch. Microbiol. 193, 77–81.

Miransari, M., 2011b. Hyperaccumulators, arbuscular mycorrhizal fungi and stress of heavy metals. Biotechnol. Adv. 29, 645–653.

Mommer, L., Kirkegaard, J., van Ruijven, J., 2016. Root–root interactions: towards a rhizosphere framework. Trends Plant Sci. 21, 1–9.

Monreal, M.A., Grant, C.A., Irvine, R.B., Mohr, R.M., McLaren, D.L., Khakbazan, M., 2011. Crop management effect on arbuscular mycorrhizae and root growth of flax. Can. J. Plant Sci. 91, 315–324.

Monzon, A., Azcon, R., 1996. Relevance of mycorrhizal fungal origin and host plant genotype to inducing growth and nutrient uptake in *Medicago* species. Agric. Ecosyst. Environ. 60, 9–15.

Mora, M., Alfaro, M., Williams, P.H., Stehr, W., Demanet, R., 2004. Effect of fertilizer input on soil acidification in relation to growth and chemical composition of a pasture and animal production. J. Soil Sci. Plant Nutr. (Chile) 4, 29–40.

Morrissey, J.P., Dow, J.M., Mark, G.L., O'Gara, F., 2004. Are microbes at the root of a solution to world food production? Rational exploitation of interactions between microbes and plants can help to transform agriculture. EMBO Rep. 5, 922–926.

Morton, J.B., 1993. Properties of infective propagules at the suborder level (Glominae versus Gigasporineae). INVAM Newsletter September, 3.

Mosier, A.R., Syers, J.K., Freney, J.R., 2004. Nitrogen fertilizer: an essential component of increased food, feed, and fiber production. In: Mosier, A.R., Syers, J.K., Freney, J.R. (Eds.), Agriculture and the Nitrogen Cycle. Assessing the Impacts of Fertilizer Use on Food Production and the Environment. SCOPE65. Island Press, Washington, DC, pp. 3–15.

Mosse, B., 1988. Some studies relating to "independent" growth of vesicular–arbuscular endophytes. Can. J. Bot. 66, 2533–2540.

Mosse, B., Hepper, C., 1975. Vesicular-arbuscular mycorrhizal infections in root organ cultures. Physiol. Plant Pathol. 5, 215–223.

Moyer–Henry, K.A., Burton, J.W., Israel, D., Rufty, T., 2006. Nitrogen transfer between plants: a $^{15}N$ natural abundance study with crop and weed species. Pant Soil 282, 7–20.

Mueller, N.D., Gerber, J.S., Johnston, M., Ray, D.K., Ramankutty, N., Foley, J.A., 2012. Closing yield gaps through nutrient and water management. Nature 490, 254–257.

Mullis, K., Faloona, F., Schaarf, S., Saiki, R., Horn, G., Erlich, H., 1986. Specific enzymatic amplification of DNA in vitro: the polymerase chain reaction. Cold Spring Harbor Symposium on Quantitative Biology 51, 263–273.

Mulugeta, D., Stoltenberg, D.E., 1997. Increased weed emergence and seed bank depletion by soil disturbance in a no-tillage system. Weed Sci. 45, 234–241.

Mummey, D.L., Rillig, M.C., 2007. Evaluation of LSU rRNA-gene PCR primers for analysis of arbuscular mycorrhizal fungal communities via terminal restriction fragment length polymorphism analysis. J. Microbiol. Methods 70, 200–204. Available from: http://dx.doi.org/10.1016/j.mimet.2007.04.002.

Munkvold, L., Kjoller, R., Vestberg, M., Rosendahl, S., Jakobsen, I., 2004. High functional diversity within species of arbuscular mycorrhizal fungi. New Phytol. 164, 357–364.

Murungu, F.S., Nyamudeza, P., Mugabe, F.T., Matimati, I., Mapfumo, S., 2006. Effects of seedling age on transplanting shock, growth and yield of pearl millet (*Pennisetum glaucum* L.) varieties in semi-arid Zimbabwe. J. Agron. 5, 205–211.

Nadeem, S.M., Ahmad, M., Zahir, Z.A., Javaid, A., Ashraf, M., 2014. The role of mycorrhizae and plant growth promoting rhizobacteria (PGPR) in improving crop productivity under stressful environments. Biotechnol. Adv. 32, 429–448.

Nakagawa, T., Okazaki, S., Shibuya, N., 2014. Genes involved in pathogenesis and defense responses. In: Tabata, S., Stougaard, J. (Eds.), The *Lotus Japonicus* Genome, Compendium of Plant Genomes. Springer, Berlin, pp. 163–169.

Nap, J.-P., Bisseling, T., 1990. Developmental biology of a plant-prokaryote symbiosis: the legume root nodule. Science, New Series 250, 948–954.

Nasim, G., 2010. The role of arbuscular mycorrhizae in inducing resistance to drought and salinity stress in crops. In: Ashraf, M., Ozturk, M., Ahmad, M.S.A. (Eds.), Plant Adaptation and Phytoremediation. Springer Science + Business Media BV, Netherlands, pp. 119–141.

Naumann, M., Schüßler, A., Bonfante, P., 2010. The obligate endobacteria of arbuscular mycorrhizal fungi are ancient heritable components related to the Mollicutes. ISME J. 4, 862–871.

Navarro, J.M., Perez-Tornero, O., Morte, A., 2014. Alleviation of salt stress in citrus seedlings inoculated with arbuscular mycorrhizal fungi depends on the rootstock salt tolerance. J. Plant. Physiol. 171, 76–85.

Neto, M.S., Scopel, E., Corbeels, M., Cardoso, A.N., Douzet, J.M., Feller, C., et al., 2010. Soil carbon stocks under no-tillage mulch-based cropping systems in the Brazilian Cerrado: an on-farm synchronic assessment. Soil Tillage Res. 110, 187–195.

Neto, M.S., Venzke Filho, S.P., Piccolo, M.C., Cerri, C.E.P., Cerri, C.C., 2009. Crop rotation under no-tillage in tibagi (Paraná State, Brazil). I Soil carbon sequestration. Revista. Brasileira de Ciência do Solo 33, 1013–1022.

Newsham, K.K., Fitter, A.H., Watkinson, A.R., 1995. Arbuscular mycorrhiza protect an annual grass from root pathogenic fungi in the field. J. Ecol. 83, 991–1000.

Ngosong, C., Gabriel, E., Ruess, L., 2014. Collembola grazing on arbuscular mycorrhiza fungi modulates nutrient allocation in plants. Pedobiologia 57, 171–179.

Ngwene, B., Gabriel, E., George, E., 2013. Influence of different mineral nitrogen sources ($NO_3^-$-N vs. $NH_4^+$-N) on arbuscular mycorrhiza development and N transfer in a *Glomus intraradices*–cowpea symbiosis. Mycorrhiza 23, 107–117.

Nicol, G.W., Leininger, S., Schleper, C., Prosser, J., 2008. The influence of soil pH on the diversity, abundance and transcriptional activity of ammonia oxidizing archaea and bacteria. Environ. Microbiol. 10, 2966–2978.

Niemi, M., Eklund, M., 1988. Effect of VA mycorrhizae and bark ash on the growth and $N_2$-fixation of two legumes. Symbiosis 6, 167–180.

Nissen, T.M., Wander, M.M., 2003. Management and soil-quality effects on fertilizer-use efficiency and leaching. Soil Sci. Soc. Am. J. 67, 1524–1532.

Nogales, A., Aguirreolea, J., María, E.S., Camprubí, A., Calvet, C., 2009. Response of mycorrhizal grapevine to Armillaria mellea inoculation: disease development and polyamines. Plant Soil 317, 177–187.

Nogueira, M.A., Cardoso, E.J.B.N., 2002. Interacções microbianas na disponibilidade e absorção de manganês por soja. Pesquisa Agropecuária Brasileira 37, 1605–1612.

Nogueira, M.A., Magalhães, G.C., Cardoso, E.J.B.N., 2004. Manganese toxicity in mycorrhizal and phosphorus-fertilized soybean plants. J. Plant Nutr. 27, 141–156.

Nogueira, M.A., Nehls, U., Hampp, R., Poralla, K., Cardoso, E.J.B.N., 2007. Mycorrhiza and soil bacteria influence extractable iron and manganese in soil and uptake by soybean. Plant Soil 298, 273–284.

Norman, J.R., Atkinson, D., Hooker, J.E., 1996. Arbuscular mycorrhizal fungal-induced alteration to root architecture in strawberry and induced resistance to the root pathogen *Phytophthora fragariae*. Plant Soil 185, 191–198.

Nurlaeny, N., Marschner, H., George, E., 1996. Effects of liming and mycorrhizal colonization on soil phosphate depletion and phosphate uptake by maize (*Zea Mays* L.) and soybean (*Glycine max* L.) grown in two tropical acid soils. Plant Soil 181, 275–285.

Oehl, F., Sieverding, E., Ineichen, K., Mäder, P., Boller, T., Wiemken, A., 2003. Impact of land use intensity on the species diversity of arbuscular mycorrhizal fungi in agroecosystems of central Europe. Appl. Environ. Microbiol. 69, 2816–2824.

Oehl, F., Sieverding, E., Ineichen, K., Ris, E.A., Boller, T., Wiemken, A., 2005. Community structure of arbuscular mycorrhizal fungi at different soil depths in extensively and intensively managed agroecosystems. New Phytol. 165, 273–283.

Oldeman, L.R., Hakkeling, R.T.A., Sombroek, W.G., 1992. World Map of the Status of Human-Induced Soil Degradation: An Explanatory Note. ISRIC, Wageningen.

Oldroyd, G.E.D., Downie, J.A., 2004. Calcium, kinases and nodulation signalling in legumes. Nat. Rev. Mol. Cell. Biol. 5, 566–576.

Oliveira, R.S., Castro, P.M.L., Dodd, J.C., Vosatka, M., 2006. Different native arbuscular mycorrhizal fungi influence the coexistence of two plant species in a highly alkaline anthropogenic sediment. Plant Soil 287, 209–221.

Oliveira, R.S., Dodd, J.C., Castro, P.M.L., 2001. The mycorrhizal status of *Phragmites australis* in several polluted soils and sediments of an industrialised region of northern Portugal. Mycorrhiza 10, 241–247.

Olsson, P.A., Johnson, N.C., 2005. Tracking carbon from the atmosphere to the rhizosphere. Ecol. Lett. 8, 1264–1270.

Öpik, M., Davison, J., Moora, M., Zobel, M., 2014. DNA-based detection and identification of Glomeromycota: the virtual taxonomy of environmental sequences 1. Botany 92, 135–147.

Öpik, M., Metsis, M., Daniell, T.J., Zobel, M., Moora, M., 2009. Large-scale parallel 454 sequencing reveals host ecological group specificity of arbuscular mycorrhizal fungi in a boreonemoral forest. New Phytol. 184, 424–437.

Öpik, M., Moora, M., Liira, J., Zobel, M., 2006. Composition of root-colonizing arbuscular mycorrhizal fungal communities in different ecosystems around the globe. J. Ecol. 94, 778–790.

Öpik, M., Moora, M., Zobel, M., Saks, U., Wheatley, R., Wright, F., et al., 2008. High diversity of arbuscular mycorrhizal fungi in a boreal herb-rich coniferous forest. New Phytol. 179, 867–876.

Öpik, M., Vanatoa, A., Vanatoa, E., Moora, M., Davison, J., Kalwij, J.M., et al., 2010. Large-scale parallel 454 sequencing reveals host ecological group specificity of arbuscular mycorrhizal fungi in a boreonemoral forest. New Phytol. 188, 223–241.

Öpik, M., Zobel, M., Cantero, J.J., Davison, J., Facelli, J.M., Hiiesalu, I., et al., 2013. Global sampling of plant roots expands the described molecular diversity of arbuscular mycorrhizal fungi. Mycorrhiza 23, 411–430.

Orchard, S., Standish, R.J., Nicol, D., Gupta, V.V.S.R., Ryan, M.H., 2016. The response of fine root endophyte (*Glomus tenue*) to waterlogging is dependent on host plant species and soil type. Plant Soil 403, 305–315.

Ortas, I., Akpinar, C., 2006. Response of kidney bean to arbuscular mycorrhizal inoculation mycorrhizal dependency in P and Zn deficient soils. Acta Agric. Scand. 56, 101–109.

Ovchinnikova, E., Journet, E.-P., Chabaud, M., Cosson, V., Ratet, P., Duc, G., et al., 2011. IPD3 controls the formation of nitrogen-fixing symbiosomes in pea and *Medicago Spp*. Mol. Plant Microbe Interact. 24, 1333–1344.

Pandey, R., Singh, B., Nair, T.V.R., 2005. Impact of arbuscular-mycorrhizal fungi on phosphorus efficiency of wheat, rye, and triticale. J. Plant. Nutr. 28, 1867–1876.

Panke-Buisse, K., Poole, A.C., Goodrich, J.K., Ley, R.E., Kao-Kniffin, J., 2015. Selection on soil microbiomes reveals reproducible impacts on plant function. ISME J. 9, 980–989.

Parry, M., Evans, A., Rosegran, M.W.T., Wheeler, T., 2009. Climate Change and Hunger: Responding to the Challenge. Published by the World Food Programme, the International Food Policy Research Institute, the New York University Center on International Cooperation, the Grantham Institute at Imperial College London, and the Walker Institute, University of Reading (United Kingdom).

Paszkowski, U., Jakovleva, L., Boller, T., 2006. Maize mutants affected at distinct stages of the arbuscular mycorrhizal symbiosis. Plant J. 47, 165–173.

Pawlowska, T.E., Taylor, J.W., 2004. Organization of genetic variation in individuals of arbuscular mycorrhizal fungi. Nature 427, 733–737. Available from: http://dx.doi.org/10.1038/nature02290.

Peay, K.G., Baraloto, C., Fine, P.V.A., 2013. Strong coupling of plant and fungal community structure across western Amazonian rainforests. ISME J. 7, 1852–1861.

Pedranzani, H., Rodríguez-Rivera, M., Gutiérrez, M., Porcel, R., Hause, B., Ruiz-Lozano, J.M., 2016. Arbuscular mycorrhizal symbiosis regulates physiology and performance of *Digitaria eriantha* plants subjected to abiotic stresses by modulating antioxidant and jasmonate levels. Mycorrhiza 26, 141–152.

Peng, S., Eissenstat, D.M., Graham, J.H., Williams, K., Hodge, N.C., 1993. Growth depression in mycorrhizal citrus at high-phosphorus supply: analysis of carbon costs. Plant Physiol. 101, 1063–1071.

Perotto, S., Brewin, N.J., Bonfante, P., 1994. Colonization of pea roots by the mycorrhizal fungus Glomus versiforme and by Rhizobium bacteria: immunological comparison using monoclonal antibodies as probes for plant cell surface components. Mol. Plant Microbe Interact. 7, 91–98.

Perry, C., Steduto, P., Allen, R.G., Burt, C.M., 2009. Increasing productivity in irrigated agriculture: agronomic constraints and hydrological realities. Agric. Water Manage. 96, 1517–1524.

Petit, E., Gubler, W.D., 2006. Influence of *Glomus intra*radices on black foot disease caused by *Cylindrocarpon macrodidymum* on *Vitis rupestris* under controlled conditions. Plant Dis. 90, 1481–1484.

Peyret-Guzzon, M., Stockinger, H., Bouffaud, M.-L., Farcy, P., Wipf, D., Redecker, D., 2015. Arbuscular mycorrhizal fungal communities and Rhizophagus irregularis populations shift in response to short-term ploughing and fertilisation in a buffer strip. Mycorrhiza 26, 33–46.

Philippot, L., Čuhel, J., Saby, N.P.A., Chèneby, D., Chroňáková, A., Bru, D., et al., 2009. Mapping field-scale spatial patterns of size and activity of the denitrifier community. Environ. Microbiol. 11, 1518–1526. Available from: http://dx.doi.org/10.1111/j.1462-2920.2009.01879.x.

Philipps, J.M., Hayman, D.S., 1970. Improved procedures for clearing roots and staining parasitic and vesicular-arbuscular mycorrhizal fungi for rapid assessment of infection. Trans. Br. Mycol. Soc. 55, 158–161.

Picone, C., 2000. Diversity and abundance of arbuscular-mycorrhizal fungus spores in tropical forest and pasture. Biotropica 32, 734–750.

Pimentel, D., Harvey, F.C., Resosudarmo, P., Sinclair, K., Kurz, D., McNair, M., et al., 1995. Environmental and economic costs of soil erosion and conservation benefits. Science, New Series 267, 1117–1123.

Pineda, A., Zheng, S.J., van Loon, J.J.A., Pieterse, C.M.J., Dicke, M., 2010. Helping plants to deal with insects: the role of beneficial soil-borne microbes. Trends Plant Sci. 15, 507–514.

Piotrowski, J.S., Denich, T., Klironomos, J.N., Graham, J.M., Rillig, M.C., 2004. The effects of arbuscular mycorrhizas on soil aggregation depend on the interaction between plant and fungal species. New Phytol. 164, 365–373.

Pivato, B., Gamalero, E., Lemanceau, P., Berta, G., 2008. Colonization of adventitious roots of *Medicago truncatula* by *Pseudomonas fluorescens* C7R12 as affected by arbuscular mycorrhiza. FEMS Microbiol. Lett. 289, 173–180.

Pivato, B., Mazurier, S., Lemanceau, P., Siblot, S., Berta, G., Mougel, C., et al., 2007. Medicago species affect the community composition of arbuscular mycorrhizal fungi associated with roots. New Phytol. 176, 197–210. Available from: http://dx.doi.org/10.1111/j.1469-8137.2007.02151.x.

Plenchette, C., Clermont-Dauphin, C., Meynard, J.M., Fortin, J.A., 2005. Managing arbuscular mycorrhizal fungi in cropping systems. Can. J. Plant Sci. 85, 31–40.

Pollock, B.J., Teplitski, M., Boinay, R.P., Bauer, W.D., Walker, G.C., 2002. A LuxR homolog controls production of symbiotically active extracellular polysaccharide II by *Sinorhizobium meliloti*. J. Bacteriol. 184, 5067–5076.

Porcel, R., Ruiz-Lozano, J.M., 2004. Arbuscular mycorrhizal influence on leaf water potential, solute accumulation, and oxidative stress in soybean plants subjected to drought stress. J. Exp. Bot. 55, 1743–1750.

Porcel, R., Aroca, R., Ruiz-Lozano, J.M., 2012. Salinity stress alleviation using arbuscular mycorrhizal fungi. A review. Agron. Sustain. Dev. 32, 181–200.

Porras-Alfaro, A., Herrera, J., Natvig, D.O., Sinsabaugh, R.L., 2007. Effect of long-term nitrogen fertilization on mycorrhizal fungi associated with a dominant grass in a semiarid grassland. Plant Soil 296, 65–75.

Porter, W.M., Abbott, L.K., Robson, A.D., 1978. Effect of rate of application of superphosphate on populations of vesicular arbuscular endophytes. Aust. J. Exp. Agric. Anim. Husb. 18, 573–578.

Poulsen, K.H., Nagy, R., Gao, L.L., Smith, S.E., Bucher, M., Smith, F.A., et al., 2005. Physiological and molecular evidence for Pi uptake via the symbiotic pathway in a reduced mycorrhizal colonization mutant in tomato associated with a compatible fungus. New Phytol. 168, 445–454.

Powell, J.R., Sikes, B.A., 2014. Method or madness: Does OTU delineation bias our perceptions of fungal ecology? New Phytol. 202, 1095–1097. Available from: http://dx.doi.org/10.1111/nph.12823.

Powell, J.R., Parrent, J.L., Hart, M.M., Klironomos, J.N., Rillig, M.C., Maherali, H., 2009. Phylogenetic trait conservatism and the evolution of functional trade-offs in arbuscular mycorrhizal fungi. Proc. R. Soc. B, Biol. Sci. 276, 4237–4245.

Power, J.F., Wilhelm, W.W., Doran, J.W., 1986. Crop residue effects on soil environment and dryland maize and soya bean production. Soil Tillage Res. 8, 101–111.

Powlson, D.S., Bhogal, A., Chambers, B.J., Colemana, K., Macdonald, A.J., Goulding, K.W.T., et al., 2012. The potential to increase soil carbon stocks through reduced tillage or organic material additions in England and Wales: a case study. Agric. Ecosyst. Environ. 146, 23–33.

Pozo, M.J., Azcón-Aguilar, C., 2007. Unraveling mycorrhiza induced resistance. Curr. Opin. Plant Biol. 10, 393–398.

Pumplin, N., Harrison, M.J., 2009. Live-cell imaging reveals periarbuscular membrane domains and organelle location in *Medicago truncatula* roots during arbuscular mycorrhizal symbiosis. Plant Physiol. 151, 809–819.

Puppi, G., Tartaglini, N., 1991. Mycorrhizal types in 3 Mediterranean communities affected by fire to different extents. Acta Oecol. Int. J. Ecol. 12, 295–304.

Püschel, D., Rydlova, J., Vosatka, M., 2007. The development of arbuscular mycorrhiza in two simulated stages of spoil-bank succession. Appl. Soil Ecol. 35, 363–369.

Qu, L.H., Michot, B., Bachellerie, J.-P., 1983. Improved methods for structure probing in large RNAs: a rapid "heterologous" sequencing approach is coupled to the direct mapping of nudease accessible sites. Application to the 5′ terminal domain of eukaryotic 28S rRNA. Nucleic Acids Res. 11, 5903–5920. Available from: http://dx.doi.org/10.1093/nar/11.17.5903.

Quast, C., Pruesse, E., Yilmaz, P., Gerken, J., Schweer, T., Yarza, P., et al., 2012. The SILVA ribosomal RNA gene database project: improved data processing and web-based tools. Nucleic Acids Res. 41, D590–D596. Available from: http://dx.doi.org/10.1093/nar/gks1219.

Quilambo, O.A., Weissenhorn, I., Doddema, H., Kuiper, P.J.C., Stulen, I., 2005. Arbuscular mycorrhizal inoculation of peanut in low-fertile tropical soil. II. Alleviation of drought stress. J. Plant Nutr. 28, 1645–1662.

Quintero-Ramos, M., Espinoza-Victoria, D., Ferrera-Cerrato, R., Bethlenfalvay, G.J., 1993. Fitting plants to soil through mycorrhizal fungi: mycorrhiza effects on plant growth and soil organic matter. Biol. Fertil. Soils 15, 103–106.

Read, D.J., 1991. Mycorrhizas in ecosystems. Experientia 47, 376–391.

Read, D.J., Birch, C.P.D., 1988. The effects and implications of disturbance of mycorrhizal mycelial systems. Proc. R. Soc. Edinburgh B., Biol. Sci. 94, 13–24.

Read, D.J., Koucheki, H.K., Hodgson, J., 1976. Vesicular–arbuscular mycorrhiza in natural vegetation systems. New Phytol. 77, 641–653.

Redecker, D., 2000. Specific PCR primers to identify arbuscular mycorrhizal fungi within colonized roots. Mycorrhiza 10, 73–80.

Redecker, D., 2002. Molecular identification and phylogeny of arbuscular mycorrhizal fungi. Plant Soil 244, 67–73.

Redecker, D., Raab, P., 2006. Phylogeny of the Glomeromycota (arbuscular mycorrhizal fungi): recent developments and new gene markers. Mycologia 98, 885–895.

Redecker, D., Morton, J.B., Bruns, T.D., 2000. Ancestral lineages of arbuscular mycorrhizal fungi (Glomales). Mol. Phylogenet. Evol. 14, 276–284. Available from: http://dx.doi.org/10.1006/mpev.1999.0713.

Redecker, D., Schussler, A., Stockinger, H., Sturmer, S.L., Morton, J.B., Walker, C., 2013. An evidence-based consensus for the classification of arbuscular mycorrhizal fungi (Glomeromycota). Mycorrhiza 23, 515–531.

Reeve, J.R., Schadt, C.W., Carpenter-Boggs, L., Kang, S., Zhou, J.Z., Reganold, J.P., 2010. Effects of soil type and farm management on soil ecological functional genes and microbial activities. ISME J. 4, 1099–1107.

Reinhardt, D., 2007. Programming good relations–development of the arbuscular mycorrhizal symbiosis. Curr. Opin. Plant. Biol. 10, 98–105.

Requena, N., Jeffries, P., Barea, J.M., 1996. Assessment of natural mycorrhizal potential in a desertified semi–arid mediterranean ecosystem. Appl. Microbiol. Biotechnol. 62, 842–847.

Requena, N., Serrano, E., Ocón, A., Breuninger, M., 1997. Plant signals and fungal perception during arbuscular mycorrhiza establishment. Phytochemistry 68, 33–40.

Rhodes, L.H., Gerdemann, J.W., 1975. Phosphate uptake zones of mycorrhizal and non-mycorrhizal onions. New Phytol. 75, 555–561.

Rich, M., Schorderet, M., Reinhardt, D., 2014. The role of the cell wall compartment in mutualistic symbioses of plants. Front. Plant Sci. 5, 238. Available from: http://dx.doi.org/10.3389/fpls.2014.00238.

Richter, D.D., Babbar, L.I., 1991. Soil diversity in the tropics. Adv. Ecol. Res. 21, 316–383.

Ricroch, A., Harwood, W., Svobodova, Z., Sági, L., Hundleby, P., Badea, E.M., et al., 2016. Challenges facing European agriculture and possible biotechnological solutions. Crit. Rev. Biotechnol. 36, 875–883. Available from: http://dx.doi.org/10.3109/07388551.2015.1055707.

Rillig, M.C., 2004a. Arbuscular mycorrhizae and terrestrial ecosystem processes. Ecol. Lett. 7, 740–754.

Rillig, M.C., 2004b. Arbuscular mycorrhizae, glomalin, and soil aggregation. Can. J. Soil Sci. 84, 355–363.

Rillig, M.C., Mummey, D.L., 2006. Mycorrhizas and soil structure. New Phytol. 171, 41–53.

Rillig, M.C., Lutgen, E.R., Ramsey, P.W., Klironomos, J.N., Gannon, J.E., 2005. Microbiota accompanying different arbuscular mycorrhizal fungal isolates influence soil aggregation. Pedobiologia 49, 251–259.

Rillig, M.C., Wright, S.F., Nichols, K.A., Schmidt, W.F., Torn, M.S., 2001. Large contribution of arbuscular mycorrhizal fungi to soil carbon pools in tropical forest soils. Plant Soil 233, 167–177.

Rivera-Becerril, F., Calantzis, C., Turnau, K., Caussanel, J.P., Belimov, A.A., Gianinazzi, S., et al., 2002. Cadmium accumulation and buffering of cadmium-induced stress by arbuscular in three *Pisum sativum* L. genotypes. J. Exp. Bot. 53, 1177–1185.

Roberts, H.A., Potter, M.E., 1980. Emergence patterns of weed seedlings in relation to cultivation and rainfall. Weed Res. 20, 377–386.

Roberts, N.J., Morieri, G., Kalsi, G., Rose, A., Stiller, J., Edwards, A., et al., 2013. Rhizobial and mycorrhizal symbioses in *Lotus japonicus* require lectin nucleotide phosphohydrolase, which acts upstream of calcium signaling. Plant Physiol. 161, 556–567.

Rodrigues, K.M., Rodrigues, B.F., 2015. Endomycorrhizal association of *Funneliformis mosseae* with transformed roots of *Linum usitatissimum*: germination, colonization, and sporulation studies. Mycology 6, 42–49.

Rodriguez-Echeverria, S., Freitas, H., 2006. Diversity of AMF associated with *Ammophila arenaria* ssp. arundinacea in Portuguese sand dunes. Mycorrhiza 16, 543–552.

Rokas, A., Williams, B.L., King, N., Carroll, S.B., 2003. Genome-scale approaches to resolving incongruence in molecular phylogenies. Nature 425, 798–804. Available from: http://dx.doi.org/10.1038/nature02053.

Rosendahl, S., 2008. Communities, populations and individuals of arbuscular mycorrhizal fungi. New Phytol. 178, 253–266.

Rosendahl, S., Stukenbrock, E.H., 2004. Community structure of arbuscular mycorrhizal fungi in undisturbed vegetation revealed by analyses of LSU rDNA sequences. Mol. Ecol. 13, 3179–3186. Available from: http://dx.doi.org/10.1111/j.1365-294X.2004.02295.x.

Rosendahl, S., McGee, P., Morton, J.B., 2009. Lack of global population genetic differentiation in the arbuscular mycorrhizal fungus *Glomus mosseae* suggests a recent range expansion which may have coincided with the spread of agriculture. Mol. Ecol. 18, 4316–4329.

Rotthauwe, J.H., Witzel, K.P., Liesack, W., 1997. The ammonia monooxygenase structural gene amoA as a functional marker: molecular fine-scale analysis of natural ammonia-oxidizing populations. Appl. Environ. Microbiol. 63, 4704–4712.

Rufyikiri, G., Declerck, S., Dufey, J.E., Delvaux, B., 2000. Arbuscular mycorrhizal fungi might alleviate aluminium toxicity in banana plants. New Phytol. 148, 343–352.

Ruiz-Lozano, J.M., 2003. Arbuscular mycorrhizal symbiosis and alleviation of osmotic stress. New perspectives for molecular studies. Mycorrhiza 13, 309–317.

Ruiz-Sánchez, M., Aroca, R., Muñoz, Y., Polón, R., Ruiz-Lozano, J.M., 2010. The arbuscular mycorrhizal symbiosis enhances the photosynthetic efficiency and the antioxidative response of rice plants subjected to drought stress. J. Plant Physiol. 167, 862–869.

Ruth, B., Khalvati, M., Schmidhalter, U., 2011. Quantification of mycorrhizal water uptake via high-resolution on-line water content sensors. Plant Soil 342, 459–468.

Ryan, M., Ash, J., 1999. Effects of phosphorus and nitrogen on growth of pasture plants and VAM fungi in SE Australian soils with contrasting fertilizer histories (conventional and biodynamic). Agric. Ecosyst. Environ. 73, 51–62.

Ryan, M.H., Angus, J.F., 2003. Arbuscular mycorrhizae in wheat and field pea crops on a low P soil: increased Zn uptake but no increase in P-uptake or yield. Plant Soil 250, 225–239.

Ryan, M.H., Ash, J.E., 1996. Colonisation of wheat in southern New South Wales by vesicular-arbuscular mycorrhizal fungi is significantly reduced by drought. Aust. J. Exp. Agric. 36, 563–569.

Ryan, M.H., Chilvers, G.A., Dumaresq, D.C., 1994. Colonisation of wheat by VA-mycorrhizal fungi was found to be higher on a farm managed in an organic manner than on a conventional neighbor. Plant Soil 160, 33–40.

Ryan, N.A., Duffy, E.M., Cassells, A.C., Jones, P.W., 2000. The effect of mycorrhizal fungi on the hatch of potato cyst nematodes. Appl. Soil Ecol. 15, 233–240.

Saito, M., Marumoto, T., 2002. Inoculation with arbuscular mycorrhizal fungi: the status quo in Japan and the future prospects. Plant Soil 244, 273–279.

Säle, V., Aguilera, P., Laczko, E., Mäder, P., Berner, A., Zihlmann, U., et al., 2015. Impact of conservation tillage and organic farming on the diversity of arbuscular mycorrhizal fungi. Soil Biol. Biochem. 84, 38–52.

Sanchez, P.A., 2002. Soil fertility and hunger in Africa. Science 295, 2019–2020.

Sanchez, L., Weidmann, S., Brechenmacher, L., Batoux, M., van Tuinen, D., Lemanceau, P., et al., 2004. Common gene expression in *Medicago truncatula* roots in response to *Pseudomonas fluorescens* colonization, mycorrhiza development and nodulation. New Phytol. 161, 855–863.

Sanders, I.R., 2002. Ecology and evolution of multigenomic arbuscular mycorrhizal fungi. Am. Nat. 160, S128–S141.

Sanders, I.R., 2004. Plant and arbuscular mycorrhizal fungal diversity—are we looking at the relevant levels of diversity and are we using the right techniques? New Phytol. 164, 415–418.

Sanders, F.E., Tinker, P.B., 1973. Phosphate flow into mycorrhizal roots. Pestic. Sci. 4, 385–395.

Sanders, F.E., Tinker, P.B., Black, R.L.B., Palmerley, S.M., 1977. The development of endomycorrhizal root systems. I. Spread of infection and growth promoting effects with four species of vesicular-arbuscular endophyte. New Phytol. 78, 257–268.

Sanders, I.R., Alt, M., Groppe, K., Boller, T., Wiemken, A., 1995. Identification of ribosomal DNA polymorphisms among and within spores of the Glomales: application to studies on the genetic diversity of arbuscular mycorrhizal fungal communities. New Phytol. 130, 419–427.

Sannazzaro, A.I., Ruiz, O.A., Alberto, E.O., Menendez, A.B., 2006. Alleviation of salt stress in *Lotus glaber* by *Glomus intraradices*. Plant Soil 285, 279–287.

Santos, J.B., Jakelaitis, A., Silva, A.A., Costa, M.D., Manabe, A., Silva, M.C.S., 2006. Action of two herbicides on the microbial activity of soil cultivated with common bean (*Phaseolus vulgaris*) in conventional-till and no-till systems. Weed Res. 46, 284–289.

Santos, J.G.D., Siqueira, J.O., Moreira, F.M.D., 2008. Efficiency of arbuscular mycorrhizal fungi isolated from bauxite mine spoils on seedling growth of native woody species. Revista Brasileira de Ciencia do Solo 32, 141–150.

Santos-Gonzalez, J.C., Finlay, R.D., Tehler, A., 2007. Seasonal dynamics of arbuscular mycorrhizal fungal communities in roots in a seminatural grassland. Appl. Environ. Microbiol. 73, 5613–5623. Available from: http://dx.doi.org/10.1128/AEM.00262-07.

Sasvár, Z., Hornok, L., Posta, K., 2011. The community structure of arbuscular mycorrhizal fungi in roots of maize grown in a 50-year monoculture. Biol. Fertil. Soils 47, 167–176.

Scheublin, T.R., van der Heijden, M.G.A., 2006. Arbuscular mycorrhizal fungi colonize nonfixing root nodules of several legume species. New Phytol. 172, 732–738.

Scheublin, T.R., Ridgway, K.P., Young, J.P.W., van der Heijden, M.G.A., 2004. Nonlegumes, legumes, and root nodules harbor different arbuscular mycorrhizal fungal communities. Appl. Environ. Microbiol. 70, 6240–6246.

Scheublin, T.R., Sanders, I.R., Keel, C., van der Meer, J.R., 2010. Characterisation of microbial communities colonising the hyphal surfaces of arbuscular mycorrhizal fungi. ISME J. 4, 752–763.

Schneider, K.D., Lynch, D.H., Dunfield, K., Khosla, K., Jansa, J., Voroney, R.P., 2015. Farm system management affects community structure of arbuscular mycorrhizal fungi. Appl. Soil Ecol. 96, 192–200. Available from: http://dx.doi.org/10.1016/j.apsoil.2015.07.015.

Schnoor, T.K., Lekberg, Y., Rosendahl, S., Olsson, P.A., 2011. Mechanical soil disturbance as a determinant of arbuscular mycorrhizal fungal communities in seminatural grassland. Mycorrhiza 21, 211–220.

Schoch, C., Seifert, K., Huhndorf, S., Robert, V., Spouge, J., Bolchacova, E., et al., 2012. Nuclear ribosomal internal transcribed spacer (ITS) region as a universal DNA barcode marker for Fungi. Proc. Natl. Acad. Sci. USA 109, 6241–6246.

Schoeneberger, M.M., Volk, R.J., Davey, C.B., 1989. Factors influencing early performance of leguminous plants in forest soils. Soil Sci. Soc. Am. J. 53, 1429–1434.

Schreiner, R.P., Koide, R.T., 1993. Mustards, mustard oils and mycorrhizas. New Phytol. 123, 107–113.

Schreiner, R.P., Mihara, K.L., McDaniel, H., Bethlenfalvay, G.J., 1997. Mycorrhizal fungi influence plant and soil functions and interactions. Plant Soil 188, 199–209.

Schüßler, A., Walker, C., 2010. The glomeromycota. A species list with new families and new genera. The Royal Botanic Garden Kew, Botanische Staatssammlung Munich, and Oregon State University.

Schüßler, A., Gehrig, H., Schwarzott, D., Walker, C., 2001a. Analysis of partial Glomales SSU rRNA gene sequences: implications for primer design and phylogeny. Mycol. Res. 105, 5–15.

Schüßler, A., Krüger, C., Urgiles, N., 2015. Phylogenetically diverse AM fungi from Ecuador strongly improve seedling growth of native potential crop trees. Mycorrhiza 26, 199–207. Available from: http://dx.doi.org/10.1007/s00572-015-0659-y.

Schüßler, A., Schwarzott, D., Walker, C., 2001b. A new fungal phylum, the Glomeromycota: phylogeny and evolution. Mycol. Res. 105, 1413–1421.

Schutte, B.J., Tomasek, B.J., Davis, A.S., Andersson, L., Benoit, D.L., Cirujeda, A., et al., 2014. An investigation to enhance understanding of the stimulation of weed seedling emergence by soil disturbance. Weed Res. 54, 1–12.

Schütz, K., Kandeler, E., Nagel, P., Scheu, S., Ruess, L., 2010. Functional microbial community response to nutrient pulses by artificial groundwater recharge practice in surface soils and subsoils. FEMS Microbiol. Ecol. 72, 445–455.

Schweiger, P.F., Jakobsen, I., 1999. Direct measurement of arbuscular mycorrhizal phosphorus uptake into field-grown winter wheat. Agron. J. 91, 998–1002.

Schweiger, R., Müller, C., 2015. Leaf metabolome in arbuscular mycorrhizal symbiosis. Curr. Opin. Plant. Biol. 26, 120–126.

Schweiger, P.F., Robson, A.D., Barrow, N.J., 1995. Root hair length determines effect of a Glomus species on shoot growth of some pasture species. New Phytol. 131, 247–254.

Seguel, A., Cumming, J.R., Klugh-Stewart, K., Cornejo, P., Borie, F., 2013. The role of arbuscular mycorrhizas in decreasing aluminium phytotoxicity in acidic soils: a review. Mycorrhiza 23, 167–183.

Selosse, M.A., Richard, F., He, X.H., Simard, S.W., 2006. Mycorrhizal networks: des liaisons dangereuses? Trends Ecol. Evol. 21, 621–628.

Selosse, M.-A., Strullu-Derrien, C., Martin, F.M., Kamoun, S., Kenrick, P., 2015. Plants, fungi and oomycetes: a 400 million year affair that shapes the biosphere? New Phytol. 206, 501–506.

Semalulu, O., Kasenge, V., Nakanwagi, J., Wagoire, W., Chemusto, S., Tukahirwa, J., 2014. Financial losses due to soil erosion in the Mt. Elgon hillsides, Uganda: a need for action. Sky J. Soil Sci. Environ. Manage. 3, 29–35.

Senés-Guerrero, C., Schüßler, A., 2016. A conserved arbuscular mycorrhizal fungal core-species community colonizes potato roots in the Andes. Fungal Diversity 77, 317–333. Available from: http://dx.doi.org/10.1007/s13225-015-0328-7.

Sheng, M., Lalande, R., Hamel, C., Ziadi, N., 2013. Effect of long-term tillage and mineral phosphorus fertilization on arbuscular mycorrhizal fungi in a humid continental zone of Eastern Canada. Plant Soil 369, 599–613.

Shi, Y., Ziadi, N., Hamel, C., Lajeunesse, J., Lafond, J., 2016. Phosphorus fertilization effect on Timothy root growth, and associated arbuscular mycorrhizal development. Agron. J. 108, 930–938.

Shipitalo, M.J., Edwards, W.M., 1998. Runoff and erosion control with conservation tillage and reduced-input practices on cropped watersheds. Soil Tillage Res. 46, 1–12.

Shirtliffe, S.J., Johnson, E.N., 2012. Progress towards no-till organic weed control in western Canada. Renew. Agric. Food Syst. 27, 60–67.

Siddiky, Md.R.K., Schaller, J., Caruso, T., Rillig, M.C., 2012. Arbuscular mycorrhizal fungi and collembola non-additively increase soil aggregation. Soil Biol. Biochem. 47, 93–99.

Sieverding, E., 1991. Vesicular-Arbuscular Mycorrhiza Management in Tropical Agrosystems. Deutsche Gesellschaft für Technische Zusammenarbeit (GTZ), Eschborn.

Sikes, B.A., 2010. When do arbuscular mycorrhizas protect plant roots from pathogens? Plant Signaling Behav. 5, 763–765.

Sikes, B.A., Cottenie, K., Klironomos, J.N., 2009. Plant and fungal identity determines pathogen protection of plant roots by arbuscular mycorrhizas. J. Ecol. 97, 1274–1280.

Sikora, R.A., Pocasangre, L., zum Felde, A., Niere, B., Vu, T.T., Dababat, A.A., 2008. Mutualistic endophytic fungi and in-planta suppressiveness to plant parasitic nematodes. Biol. Control 46, 15–23.

Simon, L., Bousquet, J., Lévesque, R.C., Lalonde, M., 1993. Origin and diversification of endomycorrhizal fungi and coincidence with vascular land plants. Nature 363, 67–69.

Simon, L., Lalonde, M., Bruns, T.D., 1992. Specific amplification of 18S fungal ribosomal genes from vesicular-arbuscular endomycorrhizal fungi colonizing roots. Appl. Environ. Microbiol. 58, 291–295.

Singh, R., Soni, S.K., Kalra, A., 2013. Synergy between *Glomus fasciculatum* and a beneficial Pseudomonas in reducing root diseases and improving yield and forskolin content in *Coleus forskohlii* Briq. under organic field conditions. Mycorrhiza 23, 35–44.

Six, J., Feller, C., Denef, K., Ogle, S., de Moraes, S.J.C., Albrecht, A., 2002. Soil organic matter, biota and aggregation in temperate and tropical soils—effects of no-tillage. Agronomie 22, 755–775.

Smit, G., Kijne, J.W., Lugtenberg, B.J.J., 1987. Involvement of both cellulose fibrils and a $Ca^{2+}$-dependent adhesin in the attachment of *Rhizobium leguminosarum* to pea root hair tips. J. Bacteriol. 169, 4294–4301.

Smith, S.E., Bowen, G.D., 1979. Soil temperature, mycorrhizal infection and nodulation of *Medicago truncatula* and *Trifolium subterraneum*. Soil Biol. Biochem. 11, 469–473.

Smith, S.E., Daft, M.J., 1977. Interactions between growth, phosphate content and nitrogen fixation in mycorrhizal and non-mycorrhizal *Medicago sativa*. Aust. J. Plant Physiol. 4, 403–413.

Smith, S.E., Read, D.J., 2008. Mycorrhizal Symbiosis, third ed. Academic Press and Elsevier, London.

Smith, S.E., Smith, F.A., 2012. Fresh perspectives on the roles of arbuscular mycorrhizal fungi in plant nutrition and growth. Mycologia 104, 1–13.

Smith, F.A., Jakobsen, I., Smith, S.E., 2000. Spatial differences in acquisition of soil phosphate between two arbuscular mycorrhizal fungi in symbiosis with *Medicago truncatula*. New Phytol. 147, 357–366.

Smith, S.E., Jakobsen, I., Grønlund, M., Smith, F.A., 2011. Roles of arbuscular mycorrhizas in plant phosphorus nutrition: interactions between pathways of phosphorus uptake in arbuscular mycorrhizal roots have important implications for understanding and manipulating plant phosphorus acquisition. Plant Physiol. 156, 1050–1057. Available from: http://dx.doi.org/10.1104/pp.111.174581.

Smith, S.E., Manjarrez, M., Stonor, R., McNeill, A., Smith, F.A., 2015. Indigenous arbuscular mycorrhizal (AM) fungi contribute to wheat phosphate uptake in a semi-arid field environment, shown by tracking with radioactive phosphorus. Appl. Soil Ecol. 96, 68–74.

Smith, S.E., Nicholas, D.J.D., Smith, F.A., 1979. Effect of early mycorrhizal infection on nodulation and nitrogen fixation in *Trifolium subterraneum* L. Aust. J. Plant. Physiol. 6, 305–311.

Smith, S.E., Smith, F.A., Jakobsen, I., 2004. Functional diversity in arbuscular mycorrhizal (AM) symbioses: the contribution of the mycorrhizal P uptake pathway is not correlated with mycorrhizal responses in growth or total P uptake. New Phytol. 162, 511–524.

Sochacki, P., Ward, J.R., Cruzan, M.B., 2013. Consequences of mycorrhizal colonization for piriqueta morphotypes under drought stress. Int. J. Plant. Sci. 174, 65–73.

Sohlenius, B., Sandor, A., 1987. Vertical distribution of nematodes in arable soil under grass (*Festuca pratensis*) and barley (*Hordeum distichum*). Biol. Fertil. Soils 3, 19–25.

Son, C.L., Smith, S.E., 1988. Mycorrhizal growth responses: interaction between photon irradiance and phosphorus nutrition. New Phytol. 108, 305–314.

Søndergaard, M., Lægaard, S., 1977. Vesicular-arbuscular mycorrhiza in some aquatic plants. Nature 268, 232–233.

Song, Y.Y., Ye, M., Li, C., He, X., Zhu–Salzman, K., Wang, R.L., et al., 2014. Hijacking common mycorrhizal networks for herbivore–induced defence signal transfer between tomato plants. Sci. Rep. 4, 3915.

Song, Y.Y., Zeng, R.S., Xu, J.F., Li, J., Shen, X., Yihdego, W.G., 2010. Interplant communication of tomato plants through underground common mycorrhizal networks. PLoS ONE 5, e13324.

Sorensen, J.N., Larse, J., Jakobsen, I., 2005. Mycorrhiza formation and nutrient concentration in leeks (*Allium porrum*) in relation to previous crop and cover crop management on high P soils. Plant Soil 273, 101–114.

Staddon, P.L., Bronk, R.C., Ostle, N., Ineson, P., Fitter, A.H., 2003. Rapid turnover of hyphae of mycorrhizal fungi determined by AMS microanalysis of $^{14}C$. Science 300, 1138–1140.

Stehouwer, R.C., Dick, W.A., Traina, S.J., 1993. Characteristics of earthworm burrow lining affecting atrazine sorption. J. Environ. Qual. 22, 181–185.

Steinberg, P.D., Rillig, M.C., 2003. Differential decomposition of arbuscular mycorrhizal fungal hyphae and glomalin. Soil Biol. Biochem. 35, 191–194.

Stockinger, H., Peyret-Guzzon, M., Koegel, S., Bouffaud, M.-L., Redecker, D., 2014. The largest subunit of RNA polymerase II as a new marker gene to study assemblages of arbuscular mycorrhizal fungi in the field. PLoS ONE 9, e107783–11. Available from: http://dx.doi.org/10.1371/journal.pone.0107783.

Stoorvogel, J.J., Smaling, E.M.A., Janssen, B.H., 1993. Calculating soil nutrient balances in Africa at different scales. I. Supra-national scale. Fert. Res. 35, 227–235.

Stribley, D.P., 1987. Mineral nutrition. In: Safir, G.R. (Ed.), Ecophysiology of VA Mycorrhizae Plants. Dep. of Botany and Plant Pathology, Michigan State University. CRC Press, Boca Raton, FL, pp. 59–70.

Sun, X.F., Su, Y.Y., Zhang, Y., Wu, M.Y., Zhang, Z., Pei, K.Q., et al., 2013. Diversity of arbuscular mycorrhizal fungal spore communities and its relations to plants under increased temperature and precipitation in a natural grassland. Chin. Sci. Bull. 58, 4109–4119.

Swanton, C.J., Murphy, S.D., 1996. Weed science beyond the weed: the role of integrated weed management (IWM) in agroecosystem health. Weed Sci. 44, 437–445.

Swanton, C.J., Weise, S.F., 1991. Integrated weed management: the rationale and approach. Weed Technol. 5, 657–663.

Sylvia, D.M., Neal, L.H., 1990. Nitrogen affects the phosphorus response of VA mycorrhiza. New Phytol. 115, 303–310.

Sylvia, D.M., Schenck, N.C.S., 1983. Application of superphosphate to mycorrhizal plants stimulates sporulation of phosphorus-tolerant vesicular-arbuscular mycorrhizal fungi. New Phytol. 95, 655–661.

Talaat, N.B., Shawky, B.T., 2011. Influence of arbuscular mycorrhizae on yield, nutrients, organic solutes, and antioxidant enzymes of two wheat cultivars under salt stress. J. Plant Nutr. Soil Sci. 174, 283–291.

Tanaka, Y., Yano, K., 2005. Nitrogen delivery to maize via mycorrhizal hyphae depends on the form of N supplied. Plant Cell Environ. 28, 1247–1254.

Tawaraya, K., Naito, M., Wagatsuma, T., 2006. Solubilization of insoluble inorganic phosphate by hyphal exudates of arbuscular mycorrhizal fungi. J. Plant Nutr. 29, 657–665.

Tawaraya, K., Saito, M., Morioka, M., Wagatsuma, T., 1996. Effect of concentration of phosphate on spore germination and hyphal growth of arbuscular mycorrhizal fungus, *Gigaspora margarita* Becker & Hall. J. Soil Sci. Plant Nutr. 42, 667–671.

Tchabi, A., Coyne, D., Hountondji, F., Lawouin, L., Wiemken, A., Oehl, F., 2010. Efficacy of indigenous arbuscular mycorrhizal fungi for promoting white yam (*Dioscorea rotundata*) growth in West Africa. Appl. Soil Ecol. 45, 92–100.

Tedersoo, L., Bahram, M., Polme, S., Koljalg, U., Yorou, N.S., Wijesundera, R., et al., 2014. Global diversity and geography of soil fungi. Science 346, 1078.

Tenkouano, A., Chantereau, J., Sereme, P., Toure, A.B., 1997. Comparative response of a day-neutral and photoperiod-sensitive sorghum to delayed sowing or transplanting. Afr. Crop Sci. J. 5, 259–266.

The Royal Society, 2009. Reaping the benefits—science and the sustainable intensification of global agriculture. Science Policy, The Royal Society, London.

Thingstrup, I., Rubaek, G., Sibbesen, E., Jakobsen, I., 1998. Flax (*Linum usitatissimum* L.) depends on arbuscular mycorrhizal fungi for growth and P uptake at intermediate but not high soil P levels in the field. Plant Soil 203, 37–46.

Thomas, R.S., Franson, R.L., Bethlenfalvay, G.J., 1993. Separation of vesicular–arbuscular mycorrhizal fungus and root effects on soil aggregation. Soil Sci. Soc. Am. J. 57, 77–81.

Thompson, J.P., 1987. Decline of vesicular–arbuscular mycorrhizae in long fallow disorder of field crops and its expression in phosphorus deficiency of sunflower. Aust. J. Agric. Res. 38, 847–867.

Thoms, C., Gattinger, A., Jacob, M., Thomas, F.M., Gleixner, G., 2010. Direct and indirect effects of tree diversity drive soil microbial diversity in temperate deciduous forest. Soil. Biol. Biochem. 42, 1558–1565.

Thomson, B.D., Robson, A.D., Abbott, L.K., 1986. Effects of phosphorus on the formation of mycorrhizas by gigaspora calospora and glomus eascicula tum in relation to root carbohydrates. New Phytol. 103, 751–765.

Thonar, C., Erb, A., Jansa, J., 2012. Real-time PCR to quantify composition of arbuscular mycorrhizal fungal communities-marker design, verification, calibration and field validation. Mol. Ecol. Res. 12, 219–232.

Thonar, C., Schnepf, A., Frossard, E., Roose, T., Jansa, J., 2011. Traits related to differences in function among three arbuscular mycorrhizal fungi. Plant Soil 339, 231–245.

Thygesen, K., Larsen, J., Bødker, L., 2004. Arbuscular mycorrhizal fungi reduce development of pea root-rot caused by Aphanomyces euteiches using oospores as pathogen inoculum. Eur. J. Plant Pathol. 110, 411–419.

Tian, C.Y., Feng, G., Li, X.L., Zhang, F.S., 2004. Different effects of arbuscular mycorrhizal fungal isolates from saline or non-saline soil on salinity tolerance of plants. Appl. Soil Ecol. 26, 143–148.

Tian, H., Drijber, R.A., Li, X., Miller, D.N., Wienhold, B.J., 2013a. Arbuscular mycorrhizal fungi differ in their ability to regulate the expression of phosphate transporters in maize (*Zea mays* L.). Mycorrhiza 23, 507–514.

Tian, H., Drijber, R.A., Zhang, J.L., Li, X.L., 2013b. Impact of long-term nitrogen fertilization and rotation with soybean on the diversity and phosphorus metabolism of indigenous arbuscular mycorrhizal fungi within the roots of maize (*Zea mays* L.). Agric. Ecosyst. Environ. 164, 53–61.

Tiemann, L.K., Grandy, A.S., Atkinson, E.E., Marin-Spiotta, E., McDaniel, M.D., 2015. Crop rotational diversity enhances belowground communities and functions in an agroecosystem. Ecol. Lett. 18, 761–771.

Tilman, D., Balzer, C., Hill, J., Befort, B.L., 2011. Global food demand and the sustainable intensification of agriculture. Proc. Natl. Acad. Sci. 108, 20260–20264.

Tilman, D., Cassman, K.G., Matson, P.A., Naylor, R., Polask, S., 2002. Agricultural sustainability and intensive production practices. Nature 418, 671–677.

Timmers, A.C.J., Auriac, M.-C., Truchet, G., 1999. Refined analysis of early symbiotic steps of the *Rhizobium-Medicago* interaction in relationship with microtubular cytoskeleton rearrangements. Development 126, 3617–3628.

Tisdall, J.M., Oades, J.M., 1982. Organic matter and water–stable aggregates in soils. J. Soil Sci. 33, 141–163.

Tisserant, E., Malbreil, M., Kuo, A., Kohler, A., Symeonidi, A., Balestrini, R., et al., 2013. Genome of an arbuscular mycorrhizal fungus provides insight into the oldest plant symbiosis. Proc. Natl. Acad. Sci. USA 110, 20117–20122. Available from: http://dx.doi.org/10.1073/pnas.1313452110.

Toljander, J.F., Artursson, V., Paul, L.R., Jansson, J.K., Finlay, R.D., 2006. Attachment of different soil bacteria to arbuscular mycorrhizal fungal extraradical hyphae is determined by hyphal vitality and fungal species. FEMS Microbiol. Lett. 254, 34–40.

Toljander, J.F., Santos-Gonzalez, J.C., Tehler, A., Finlay, R.D., 2008. Community analysis of arbuscular mycorrhizal fungi and bacteria in the maize mycorrhizosphere in a long-term fertilization trial. FEMS Microbiol. Ecol. 65, 323–338.

Tommerup, I.C., 1984. Persistence of infectivity by germinated spores of vesicular–arbuscular mycorrhizal fungi in soil. Trans. Br. Mycol. Soc. 82, 275–282.

Tommerup, I.C., Abbott, L.K., 1981. Prolonged survival and viability of VA mycorrhizal hyphae after root death. Soil Biol. Biochem. 13, 431–433.

Torrecillas, E., Alguacil, M.M., Roldán, A., 2012. Host preferences of arbuscular mycorrhizal fungi colonizing annual herbaceous plant species in semiarid Mediterranean prairies. Appl. Environ. Microbiol. 78, 6180–6186.

Torres-Barragán, A., Zavaleta-Mejía, E., González-Chávez, C., Ferrera-Cerrato, R., 1996. The use of arbuscular mycorrhizae to control onion white rot (*Sclerotium cepivorum* Berk.) under field conditions. Mycorrhiza 6, 253–257.

Towery D., 1998. No-tilľs impact on water quality. In: Sixth Argentine National Congress of Direct Drilling (in Spanish AAPRESID) Mar de Plata, Argentina, pp. 17–26.

Trappe, J.M., Molina, R., Castellano, M., 1984. Reactions of mycorrhizal fungi and mycorrhiza formation to pesticides. Annu. Rev. Phytopathol. 22, 331–359.

Troeh, Z.I., Loynachan, T.E., 2003. Endomycorrhizal fungal survival in continuous corn, soybean, and fallow. Agron. J. 95, 224–230.

Trouvelot, A., Kough, J.L., Gianinazzi-Pearson, V., 1986. Mesure du taux de mycorhization VA dun système radiculaire. Recherche de méthodes destimation ayant une signification fonctionnelle. In: Gianinazzi-Pearson, V., Gianinazzi, S. (Eds.), Physiological and Genetical Aspects of Mycorrhiza. INRA presse, Paris, pp. 217–221.

Trouvelot, S., van Tuinen, D., Hijri, M., Gianinazzi-Pearson, V., 1999. Visualization of DNA loci in interphasic nuclei of glomalean fungi by fluorescence in situ hybridization. Mycorrhiza 8, 203–208.

Tsiafouli, M.A., Thebault, E., Sgardelis, S.P., de Ruiter, P.C., van der Putten, W.H., Birkhofer, K., et al., 2015. Intensive agriculture reduces soil biodiversity across Europe. Global Change Biol. 21, 973–985.

Tuck, S.L., Winqvist, C., Mota, F., Ahnstrom, J., Turnbull, L.A., Bengtsson, J., 2014. Land-use intensity and the effects of organic farming on biodiversity: a hierarchical meta-analysis. J. Appl. Ecol. 51, 746–755.

Turnau, K., Haselwandter, K., 2002. Arbuscular mycorrhizal fungi, an essential component of soil microflora in ecosystem restoration. In: Gianinazzi, S., Schüepp, H., Barea, J.M.,

Haselwandter, K. (Eds.), Mycorrhizal Technology in Agriculture: From Genes to Bioproducts. Birkhäuser Basel, Basel, pp. 137–149.

Turnau, K., Ryszka, P., Gianinazzi-Pearson, V., van Tuinen, D., 2001. Identification of arbuscular mycorrhizal fungi in soils and roots of plants colonizing zinc wastes in southern Poland. Mycorrhiza 10, 169–174.

Turrini, A., Giordani, T., Avio, L., Natali, L., Giovannetti, M., Cavallini, A., 2016. Large variation in mycorrhizal colonization among wild accessions, cultivars, and inbreds of sunflower (*Helianthus annuus* L.). Euphytica 207, 331–342.

Uri, N.D., 1999. Factors affecting the use of conservation tillage in the United States. Water Air Soil Pollut. 116, 621–638.

US Census Office, 2014. World population. Available at: <http://www.census.gov/population/international/data/idb/informationGateway.php> (accessed 03.12.15.).

van Brussel, A.A.N., Bakhuizen, R., van Spronsen, P.C., Spaink, H.P., Tak, T., Lugtenberg, B.J.J., et al., 1992. Induction of pre-infection thread structures in the leguminous host plant by mitogenic lipo-oligosaccharides of Rhizobium. Science 257, 70–72.

Vance, C.P., 2001. Symbiotic nitrogen fixation and phosphorus acquisition. Plant nutrition in a world of declining renewable resources. Plant Physiol. 127, 390–397.

Vandenkoornhuyse, P.J., Mahé, S., Ineson, P., Staddon, P., Ostle, N., Cliquet, J.B., et al., 2007. Active root-inhabiting microbes identified by rapid incorporation of plant-derived carbon into RNA. PNAS 104, 16070–16975.

Van de Peer, Y., Chapelle, S., De Wachter, R., 1996. A quantitative map of nucleotide substitution rates in bacterial rRNA. Nucleic Acids Res. 24, 3381–3391.

van der Heijden, M.G.A., 2010. Mycorrhizal fungi reduce nutrient loss from model grassland ecosystems. Ecology 91, 1163–1171.

van der Heijden, M.G.A., Horton, T.R., 2009. Socialism in soil? The importance of mycorrhizal fungal networks for facilitation in natural ecosystems. J. Ecol. 97, 1139–1150.

van der Heijden, M.G.A., Boller, T., Wiemken, A., Sanders, I.R., 1998a. Different arbuscular mycorrhizal fungal species are potential determinants of plant community structure. Ecology 79, 2082–2091.

van der Heijden, M.G.A., Klironomos, J.N., Ursic, M., Moutoglis, P., Streitwolf-Engel, R., Thomas Boller, T., et al., 1998b. Mycorrhizal fungal diversity determines plant biodiversity, ecosystem variability and productivity. Nature 396, 69–72.

van der Heijden, M.G.A., Martin, F.M., Selosse, M.A., Sanders, I.R., 2015. Mycorrhizal ecology and evolution: the past, the present, and the future. New Phytol. 205, 1406–1423.

van der Heijden, M.G.A., Scheublin, T.R., Brader, A., 2004. Taxonomic and functional diversity in arbuscular mycorrhizal fungi—is there any relationship? New Phytol. 164, 201–204.

van der Heijden, M.G.A., Wiemken, A., Sanders, I.R., 2003. Different arbuscular mycorrhizal fungi alter coexistence and resource distribution between co-occurring plant. New Phytol. 157, 569–578.

van der Heijden, M.G.A., Bardgett, R.D., van Straalen, N.M., 2008a. The unseen majority: soil microbes as drivers of plant diversity and productivity in terrestrial ecosystems. Ecol. Lett. 11, 296–310.

van der Heijden, M.G.A., Verkade, S., de Bruin, S.J., 2008b. Mycorrhizal fungi reduce the negative effects of nitrogen enrichment on plant community structure in dune grassland. Global Change Biology 14, 2626–2635.

Van Geel, M., Busschaert, P., Honnay, O., Lievens, B., 2014. Evaluation of six primer pairs targeting the nuclear rRNA operon for characterization of arbuscular mycorrhizal fungal

(AMF) communities using 454 pyrosequencing. J. Microbiol. Methods 106, 93–100. Available from: http://dx.doi.org/10.1016/j.mimet.2014.08.006.

van Kessel, C., Singleton, P.W., Hoben, H.J., 1985. Enhanced N-transfer from a soybean to maize by vesicular arbuscular mycorrhizal (VAM) fungi. Plant Physiol. 79, 562–563.

van Tuinen, D., Jacquot, E., Zhao, B., Gollotte, A., Gianinazzi-Pearson, V., 1998a. Characterization of root colonization profiles by a microcosm community of arbuscular mycorrhizal fungi using 25S rDNA-targeted nested PCR. Mol. Ecol. 7, 879–887.

van Tuinen, D., Zhao, B., Gianinazzi-Pearson, V., 1998b. PCR in studies of AM fungi: from primers to application. In: Varma, A. (Ed.), Mycorrhiza Manual. Springer, Heidelberg, pp. 387–400.

Van Vuuren, D.P., Bouwman, A.F., Beusen, A.H.W., 2010. Phosphorus demand for the 1970–2100 period: a scenario analysis of resource depletion. Global Environ. Chang. 20, 428–439.

Varela-Cervero, S., Vasar, M., Davison, J., Barea, J.M., Opik, M., Azcón-Aguilar, C., 2015. The composition of arbuscular mycorrhizal fungal communities differs among the roots, spores and extraradical mycelia associated with five *Mediterranean* plant species. Environ. Microbiol. 17, 2882–2895.

Varela-Cervero, S., López-García, Á., Barea, J.M., Azcón-Aguilar, C., 2016a. Spring to autumn changes in the arbuscular mycorrhizal fungal community composition in the different propagule types associated to a *Mediterranean* shrubland. Plant Soil 408 (1–2), 107–120.

Varela-Cervero, S., López-García, Á., Barea, J.M., Azcón-Aguilar, C., 2016b. Differences in the composition of arbuscular mycorrhizal fungal communities promoted by different propagule forms from a Mediterranean shrubland. Mycorrhiza 26, 489–496.

Venke Filho, S.P., Feigl, B.J., Sá, J.C.M., Clemente, C., 1999. Colonização por fungos micorrízicos arbusculares em milho e soja em uma cronosequência the sistema de plantio directo (Arbuscular colonization of maize and soybean in a crono-sequence of no-till system). Anais do XXVII Congresso Brasileiro de Ciência do Solo. Revista de Plantio Directo 54, 34.

Verbruggen, E., Kiers, E.T., 2010. Evolutionary ecology of mycorrhizal functional diversity in agricultural systems. Evol. Appl. 3, 547–560.

Verbruggen, E., Roling, W.F.M., Gamper, H.A., Kowalchuk, G.A., Verhoef, H.A., van der Heijden, M.G.A., 2010. Positive effects of organic farming on below-ground mutualists: large-scale comparison of mycorrhizal fungal communities in agricultural soils. New Phytol. 186, 968–979.

Verbruggen, E., Van Der Heijden, M.G.A., Weedon, J.T., Kowalchuk, G.A., Röling, W.F.M., 2012. Community assembly, species richness and nestedness of arbuscular mycorrhizal fungi in agricultural soils. Mol. Ecol. 21, 2341–2353.

Veresoglou, S.D., Chen, B., Rillig, M.C., 2012. Arbuscular mycorrhiza and soil nitrogen cycling. Soil Biol. Biochem. 46, 53–62.

Vesper, S.J., Bauer, W.D., 1986. Role of pili (fimbriae) in attachment of *Bradyrhizobium japonicum* to soybean roots. Appl. Environ. Microbiol. 52, 134–141.

Vessey, J.K., Pawlowski, K., Bergman, B., 2004. Root-based $N_2$-fixing symbioses: legumes, actinorhizal plants, *Parasponia* sp. and cycads. Plant Soil 266, 205–230.

Vestberg, M., Saari, K., Kukkonen, S., Hurme, T., 2005. Mycotrophy of crops in rotation and soil amendment with peat influence the abundance and effectiveness of indigenous arbuscular mycorrhizal fungi in field soil. Mycorrhiza 15, 447–458.

Vieira, F.C.S., Nahas, E., 2005. Comparison of microbial numbers in soils by using various culture media and temperatures. Microbiol. Res. 160, 197–202.

Vierheilig, H., 2004. Further root colonization by arbuscular mycorrhizal fungi in already mycorrhizal plants is suppressed after a critical level of root colonization. J. Plant Physiol. 161, 339–341.

Voets, L., Goubau, I., Olsson, P.A., Merckx, R., Declerck, S., 2008. Absence of carbon transfer between *Medicago truncatula* plants linked by a mycorrhizal network, demonstrated in an experimental microcosm. FEMS Microbiol. Ecol. 65, 350–360.

Voets, L., Providencia, I.E., Fernandez, K., IJdo, M., Cranenbrouck, S., Declerck, S., 2009. Extraradical mycelium network of arbuscular mycorrhizal fungi allows fast colonization of seedlings under in vitro conditions. Mycorrhiza 19, 347–356.

Vogelsang, K.M., Reynolds, H.L., Bever, J.D., 2006. Mycorrhizal fungal identity and richness determine the diversity and productivity of a tallgrass prairie system. New Physiol. 172, 554–562.

Vukicevich, E., Lowery, T., Bowen, P., Úrbez-Torres, J.R., Hart, M., 2016. Cover crops to increase soil microbial diversity and mitigate decline in perennial agriculture. A review. Agron. Sustain. Dev. 36, 48. Available from: http://dx.doi.org/10.1007/s13593-016-0385-7.

Vuksani, A., Sallaku, G., Balliu, A., 2015. The effects of endogenous mycorrhiza (*Glomus* spp.) on stand establishment rate and yield of open field tomato crop. Albanian J. Agric. Sci. 14, 25–30.

Wagg, C., Bender, S.F., Widmer, F., van der Heijden, M.G.A., 2014. Soil biodiversity and soil community composition determine ecosystem multifunctionality. Proc. Natl. Acad. Sci. USA 111, 5266–5270.

Waggoner, P.E., 1995. How much land can ten billion people spare for nature? Does technology make a difference? Technol. Soc. 17, 17–34.

Walder, F., Niemann, H., Natarajan, M., Lehmann, M.F., Boller, T., Wiemken, A., 2012. Mycorrhizal networks: common goods of plants shared under unequal terms of trade. Plant Physiol. 159, 789–797.

Walker, C., Schüßler, A., 2004. Nomenclatural clarifications and new taxa in the glomeromycota. Mycol. Res. 108, 981–982.

Wall, D.H., Bardgett, R.D., Kelly, E.F., 2010. Biodiversity in the dark. Nat. Geosci. 3, 297–298.

Walley, F.L., Germida, J.J., 1995. Estimating the viability of vesicular arbuscular mycorrhizae fungal spores using tetrazolium salts as vital stains. Mycologia 87, 273–279.

Wang, B., Qju, Y.L., 2006. Phylogenetic distribution and evolution of mycorrhizas in land plants. Mycorrhiza 16, 299–363.

Wang, H., Li, H.B., 2013. Study on in-situ bioremediation of polycyclic aromatic hydrocarbon contaminated farmland soil. Adv. Mater. Res. 610–613, 1359–1363.

Wang, M.Y., Hu, L.B., Wang, W.H., Liu, S.T., Li, M., Liu, R.J., 2009. Influence of long-term fixed fertilization on diversity of arbuscular mycorrhizal fungi. Pedosphere 19, 663–672.

Wang, P., Shu, B., Wang, Y., Zhang, D.J., Liu, J.F., Xia, R.X., 2013. Diversity of arbuscular mycorrhizal fungi in red tangerine (*Citrus reticulate* Blanco) rootstock rhizospheric soils from hillside citrus orchards. Pedobiologia 56, 161–167.

Warnock, D.D., Lehmann, J., Kuyper, T.W., Rillig, M.C., 2007. Mycorrhizal responses to biochar in soil—concepts and mechanisms. Plant Soil 300, 9–20.

Wehner, J., Antunes, P.M., Powell, J.R., Mazukatow, J., Rillig, M.C., 2010. Plant pathogen protection by arbuscular mycorrhizas: a role for fungal diversity? Pedobiologia 53, 197–201.

Whipps, J.M., 2004. Prospects and limitations for mycorrhizas in biocontrol of root pathogens. Can. J. Bot. 82, 1198–1227.

White, K.D., 1970. Fallowing, crop rotation, and crop yields in roman times. Agric. Hist. 44, 281–290.

Whittingham, J., Read, D.J., 1982. Vesicular–arbuscular mycorrhiza in natural vegetation systems. III. Nutrient transfer between plants with mycorrhizal interconnections. New Phytol. 90, 277–284.

Woese, C.R., 1987. Bacterial evolution. Microbiol. Rev. 51, 221–271.

Wolfe, B.E., Mummey, D.L., Rillig, M.C., Klironomos, J.N., 2006. Small-scale spatial heterogeneity of arbuscular mycorrhizal fungal abundance and community composition in a wetland plant community. Mycorrhiza 327, 175–183. Available from: http://dx.doi.org/10.1007/s00572-006-0089-y.

Wright, S.F., Upadhyaya, A., 1996. Extraction of an abundant and unusual protein from soil and comparison with hyphal protein of arbuscular mycorrhizal fungi. Soil Sci. 161, 575–586.

Wright, S.F., Franke-Snyder, M., Morton, J.B., Upadhyaya, A., 1996. Time-course study and partial characterization of a protein on hyphae of arbuscular mycorrhizal fungi during active colonization of roots. Plant Soil 181, 193–203.

Wu, Q.S., Srivastava, A.K., Zou, Y.N., 2013. AMF-induced tolerance to drought stress in citrus: a review. Sci. Hortic. (Amsterdam) 164, 77–87.

Xavier, L.J.C., Germida, J.J., 2003. Bacteria associated with *Glomus clarum* spores influence mycorrhizal activity. Soil Biol. Biochem. 35, 471–478.

Xie, X., Yoneyama, K., Yoneyama, K., 2010. The strigolactone story. Annu. Rev. Phytopathol. 4, 93–117.

Yang, H., Koide, R.T., Zhang, Q., 2016. Short-term waterlogging increases arbuscular mycorrhizal fungal species richness and shifts community composition. Plant Soil 404, 373–384.

Yang, W.-C., Katinakis, P., Hendriks, P., Smolders, A., de Vries, F., Spee, J., et al., 1993. Characterization of Gm ENOD40, a gene showing novel patterns of cell-specific expression during soybean nodule development. Plant J. 3, 573–585.

Yang, Y., Song, Y., Scheller, H.V., Ghosh, A., Ban, Y., Chen, H., et al., 2015. Community structure of arbuscular mycorrhizal fungi associated with *Robinia pseudoacacia* in uncontaminated and heavy metal contaminated soils. Soil Biol. Biochem. 86, 146–158.

Yano, K., Takaki, M., 2005. Mycorrhizal alleviation of acid soil stress in the sweet potato (*Ipomoea batatas*). Soil Biol. Biochem. 37, 1569–1572.

Young, N.D., Debelle, F., Oldroyd, G.E., Geurts, R., Cannon, S.B., Udvardi, M.K., et al., 2011. The *Medicago* genome provides insight into the evolution of rhizobial symbioses. Nature 480, 22–29.

Zaller, J.G., Heigl, F., Ruess, L., Grabmaier, A., 2014. Glyphosate herbicide affects belowground interactions between earthworms and symbiotic mycorrhizal fungi in a model ecosystem. Sci. Rep. 4, 5634. Available from: http://dx.doi.org/10.1038/srep05634.

Zangaro, W., Rostirola, L.V., de Souza, P.B., Alves, R.D., Lescano, L., Rondina, A.B.L., et al., 2013. Root colonization and spore abundance of arbuscular mycorrhizal fungi in distinct successional stages from an Atlantic rainforest biome in southern Brazil. Mycorrhiza 23, 221–233.

Zhang, J., Subramanian, S., Stacey, G., Yu, O., 2009. Flavones and flavonols play distinct critical roles during nodulation of *Medicago truncatula* by *Sinorhizobium meliloti*. Plant J. 57, 171–183.

Zhang, W.J., Jiang, F.B., Ou, J.F., 2011. Global pesticide consumption and pollution: with China as a focus. Proc. Int. Acad. Ecol. Environ. Sci. 1, 125–144.

Zhu, X.C., Song, F.B., Liu, S.Q., Liu, T.D., Zhou, X., 2012. Arbuscular mycorrhizae improves photosynthesis and water status of *Zea mays* L. under drought stress. Plant Soil Environ. 58, 186–191.

# Index

*Note*: Page numbers followed by "*f*," "*t*," and "*b*" refer to figures, tables, and boxes, respectively.

## A

## B

## M



## N

Printed in the United States
By Bookmasters